**Springer Theses**

Recognizing Outstanding Ph.D. Research

## Aims and Scope

The series "Springer Theses" brings together a selection of the very best Ph.D. theses from around the world and across the physical sciences. Nominated and endorsed by two recognized specialists, each published volume has been selected for its scientific excellence and the high impact of its contents for the pertinent field of research. For greater accessibility to non-specialists, the published versions include an extended introduction, as well as a foreword by the student's supervisor explaining the special relevance of the work for the field. As a whole, the series will provide a valuable resource both for newcomers to the research fields described, and for other scientists seeking detailed background information on special questions. Finally, it provides an accredited documentation of the valuable contributions made by today's younger generation of scientists.

## Theses are accepted into the series by invited nomination only and must fulfill all of the following criteria

- They must be written in good English.
- The topic should fall within the confines of Chemistry, Physics, Earth Sciences, Engineering and related interdisciplinary fields such as Materials, Nanoscience, Chemical Engineering, Complex Systems and Biophysics.
- The work reported in the thesis must represent a significant scientific advance.
- If the thesis includes previously published material, permission to reproduce this must be gained from the respective copyright holder.
- They must have been examined and passed during the 12 months prior to nomination.
- Each thesis should include a foreword by the supervisor outlining the significance of its content.
- The theses should have a clearly defined structure including an introduction accessible to scientists not expert in that particular field.

More information about this series at http://www.springer.com/series/8790

Catherine E. Scott

# The Biogeochemical Impacts of Forests and the Implications for Climate Change Mitigation

Doctoral Thesis accepted by
the University of Leeds, UK

*Author*
Dr. Catherine E. Scott
School of Earth and Environment
University of Leeds
Leeds
UK

*Supervisors*
Dr. Dominick Spracklen
Prof. Piers Forster
Prof. Kenneth Carslaw
School of Earth and Environment
University of Leeds
Leeds
UK

ISSN 2190-5053 ISSN 2190-5061 (electronic)
ISBN 978-3-319-36214-4 ISBN 978-3-319-07851-9 (eBook)
DOI 10.1007/978-3-319-07851-9

Springer Cham Heidelberg New York Dordrecht London


Softcover reprint of the hardcover 1st edition 2014


Printed on acid-free paper

Springer is part of Springer Science+Business Media (www.springer.com)

**Parts of this thesis have been published in the following journal articles:**

C. E. Scott, A. Rap, D. V. Spracklen, P. M. Forster, K. S. Carslaw, G. W. Mann, K. J. Pringle, N. Kivekäs, M. Kulmala, H. Lihavainen and P. Tunved (2014) "The Direct and Indirect Radiative Effects of Biogenic Secondary Organic Aerosol", *Atmospheric Chemistry and Physics*, **14**, 447–470.

F. Riccobono, S. Schobesberger, C. E. Scott *et al.*, (2014) "Oxidation Products of Biogenic Emissions Contribute to Nucleation of Atmospheric Particles", *Science,* **344** (6185), 717–721.

C. E. Scott, D. V. Spracklen, J. R. Pierce, I. Riipinen, S. D. D'Andrea, A. Rap, K. S. Carslaw, P. M. Forster, M. Kulmala, G. W. Mann, K. Pringle (2014) "Impact of partitioning approach on the indirect radiative effect of biogenic secondary organic aerosol", (in prep).

# Supervisor's Foreword

Forests and vegetation emit biogenic volatile organic compounds (BVOCs) into the atmosphere which, once oxidised, can partition into the particle phase, forming secondary organic aerosol (SOA). This thesis reports a unique and comprehensive analysis of the impact of BVOC emissions on atmospheric aerosol and climate. A state-of-the-art global aerosol microphysics model is used to make the first detailed assessment of the impact of BVOC emissions on aerosol microphysical properties, improving our understanding of the role of these emissions in controlling the Earth's climate.

Laboratory experiments, ambient atmospheric observations and model simulations are combined to provide new evidence that the oxidation products of BVOCs participate in the first steps of particle formation. This finding means that BVOCs have a larger impact on climate than previously thought, with the global radiative effect of biogenic SOA estimated to be half the net anthropogenic radiative forcing of climate. The thesis also reports on the implications for the climate impact of forests. Accounting for the climate impacts of SOA, alongside the carbon cycle and surface albedo effects that have been studied in previous work, increases the total warming effect of global deforestation by 21 %. The thesis suggests that deforestation warms climate more than previously thought.

Leeds, UK, June 2014 Dr. Dominick Spracklen

# Acknowledgments

Firstly I would like to thank my supervisors: Piers Forster, Dom Spracklen and Ken Carslaw. Without their guidance, advice and support over the past four years, this work, and the production of this thesis would not have been possible. I am extremely grateful for the opportunities they have given me, and the enthusiasm with which they have provided supervision.

Secondly, I would like to thank Alexandru Rap for calculating the direct radiative forcings presented in Chap. 4, Jeff Pierce for conducting the TOMAS simulations discussed in Chap. 5 and Steve Arnold for running the Community Land Model used in Chap. 6. I would also like to thank Francesco Riccobono and the "CLOUD" team for providing the mechanism for new particle formation described in Chap. 3. For providing the measurement data used in Chap. 3, I would like to thank: Markku Kulmala, Pasi Aalto, Peter Tunved, Niku Kivekas and Heikki Lihavainen. I would also like to thank Ilona Riipinen and Steve D'Andrea with whom I have collaborated. Also thanks to Richard Rigby, without whose tireless IT support, this work would definitely not have been possible!

During the course of my Ph.D. I have very much valued being part of three great research groups: The Physical Climate Change group, the Aerosol Modelling group and the Biosphere-Atmosphere Group, and would like to thank all members for making me feel very welcome. Deserving of a particular mention for their help and advice over the years are: Anja, Carly, Annabel, Eimear, Alex, Matt, Helen, Tom, Kirsty, Graham, Lindsay, Jo, Steve P., Lawrence, Jayne, Julia, Neil, Susi, Sarah L., Ros, Mike and Sarah M.

For supporting me over the past four years, I would like to thank my Mum, Dad and Sally; Chris, Adèle and Wayne (and Myla!); Erin, Zarashpe, Ed, Tim, Rosey, Susie, Louisa and Al; Vicky, Amy, my fellow 2009–2013 Ph.D. gang: Amber, Jenn, Dave, Bradley and James; and last but definitely not least, thank you to Ross for everything!

# Contents

# Abbreviations and Acronyms

| | |
|---|---|
| AIE | Aerosol Indirect Effect |
| AOGCM | Atmosphere–Ocean General Circulation Model |
| BHN | Binary Homogeneous Nucleation |
| BLN | Boundary Layer Nucleation |
| BVOC | Biogenic Volatile Organic Compound |
| CCN | Cloud Condensation Nuclei |
| CDNC | Cloud Droplet Number Concentration |
| CLM | Community Land Model |
| $CO_2e$ | Carbon Dioxide Equivalent |
| DRE | Direct Radiative Effect |
| ECS | Equilibrium Climate Sensitivity |
| E–S | Edwards–Slingo Radiative Transfer Model |
| ESM | Earth System Model |
| FAO | Food and Agriculture Organisation of the United Nations |
| GCM | General Circulation Model |
| GFED | Global Fire Emissions Database |
| GHG | Greenhouse Gas |
| GLOMAP | Global Model of Aerosol Processes |
| IPCC | Intergovernmental Panel on Climate Change |
| LW | Longwave |
| MEGAN | Model of Emissions of Gases and Aerosol from Nature |
| ppbv | Parts per billion by volume |
| pptv | Parts per trillion by volume |
| RCP | Representative Concentration Pathway |
| RE | Radiative Effect |
| RF | Radiative Forcing |
| SMPS | Scanning Mobility Particle Sizer |
| SOA | Secondary Organic Aerosol |
| SRES | Special Report on Emission Scenarios |
| SS | Supersaturation |
| $SS_{max}$ | Maximum Supersaturation |

| | |
|---|---|
| SW | Shortwave |
| TOA | Top of the Atmosphere |
| UNFCCC | United Nations Framework Convention on Climate Change |

# Chapter 1
# Introduction

Vegetation emits biogenic volatile organic compounds (BVOCs) into the atmosphere which, once oxidised, may partition into the particle-phase forming secondary organic aerosol (SOA). In this thesis, the climatic impacts of biogenic SOA are quantified, using a detailed global aerosol microphysics model, and the sensitivity of these radiative effects to the representation of various atmospheric processes is examined.

The radiative effects of biogenic SOA have implications for the climatic impact of forests and any changes to their distribution. In this thesis, simple deforestation scenarios are used to quantify the radiative effects of potential changes to the magnitude and distribution of biogenic SOA production.

## 1.1 The Climate System

The energy balance of the Earth system is governed by fluxes of incoming and outgoing radiation. Shortwave (SW) solar radiation reaching the Earth system is either reflected or absorbed, as dictated by the nature of the surface and composition of the atmosphere. Figure 1.1 depicts the estimated global average energy flows within the Earth system and indicates that the atmosphere exerts considerable control over how much incoming radiation reaches the surface.

Almost half of the incoming solar radiation is absorbed at the Earth's surface and re-emitted at longer wavelengths (LW); the Earth system behaves almost as a black body with a total emissive power ($\sigma T^4$) proportional to the fourth power of its temperature, $T$, where $\sigma$ is the Stefan-Boltzmann constant. In its equilibrium state, the amount of energy entering the Earth system is equal to that which leaves. If LW radiation is prevented from leaving, the system shifts in order to maintain the required outgoing flux, with an increase in emissive temperature. As such, changes to atmospheric composition, levels of incoming SW radiation, and



C.E. Scott, *The Biogeochemical Impacts of Forests and the Implications for Climate Change Mitigation*, Springer Theses, DOI 10.1007/978-3-319-07851-9_1

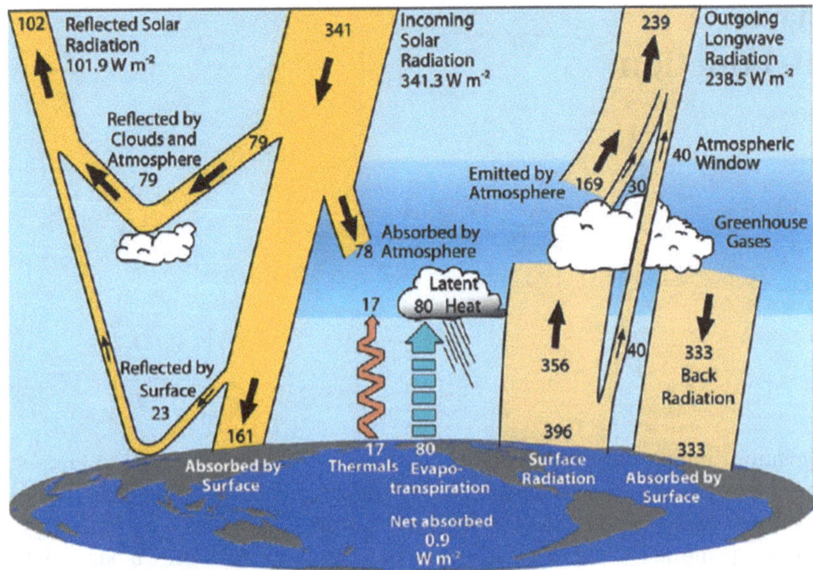


**Fig. 1.1** Schematic of the estimated global average energy budget (values in W $m^{-2}$) for March 2000 to May 2004; reproduced from Trenberth et al. [225]

properties of the Earth's surface that initiate a radiative imbalance, can alter the temperature.

A radiative forcing (RF) may be assigned to physical factors on the basis of their ability to perturb the Earth's radiative balance, over a given time period. The RF is most commonly defined as the net change in radiative flux (i.e. downwards minus upwards) at the tropopause, after allowing for stratospheric temperatures to readjust to radiative equilibrium but with surface and tropospheric temperatures and state held fixed at their unperturbed values [58, 82, 189]. As the surface and troposphere are strongly coupled by convective heat transfer, there is a direct relationship between the RF across the tropopause and the global mean equilibrium temperature change at the Earth's surface ($\Delta T_s$), such that $\Delta T_s = \lambda$ RF [189], where $\lambda$ represents the *climate sensitivity parameter* (in K $W^{-1}$ $m^2$). However, the relationship between RF and transient climate change is not straightforward, and consideration must be given to the temporal and spatial variability of the forcing agent.

## 1.2 Forests and Their Impact on the Climate

Approximately 31 % of the Earth's land area (4.03 billion hectares) is presently covered by forest (Fig. 1.2); one third of this being undisturbed primary forest [55]. More than half of the world's current forest cover lies within just five countries: The Russian Federation, Brazil, Canada, the United States of America and China [55].

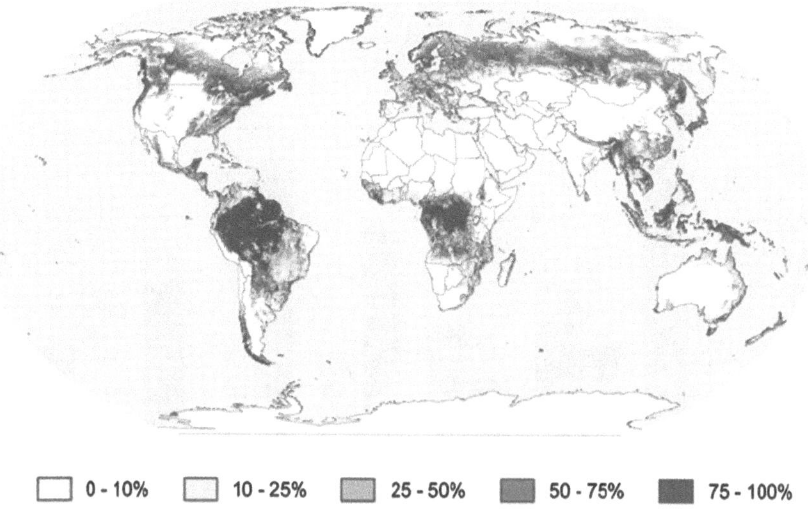


**Fig. 1.2** Estimated percentage forest cover in 2000; reproduced from Hansen et al. [83]

## *1.2.1 Historical Changes to Forest Distribution*

Land-use change has accompanied population growth for many years, particularly the last 6000. As indicated in Fig. 1.3, prior to 1850 deforestation occurred mostly in temperate regions (Asia, Europe and North America). In medieval Europe, deforestation in France, Germany and the UK occurred mainly on the land considered most suitable for farming; the extensive coniferous forests of Finland, Norway and Sweden escaped relatively unscathed during this time, with deforestation in these countries concentrated around cities and less widespread than in other European countries [244].

From around 1900 onwards, the majority of deforestation has occurred in tropical regions (Fig. 1.3), specifically South and Central America, South-east Asia and Central Africa; during the 1990s approximately 16 million hectares of forest was lost globally per year. As with temperate deforestation, forest clearance in the tropics occurs predominantly to acquire land suitable for agriculture [63], in particular for food and fuel crop growth in Africa and South-east Asia [30, 62, 134, 142], and cattle ranching in South and Central America [57, 66]. Pressure on the land from mineral mining (e.g. to obtain gold, copper, tin), coal mining, and oil drilling is also particularly prevalent in parts of South America, Africa and South-east Asia [66].

Since 2000, the rate of tropical forest loss has slowed (Fig. 1.4 and e.g. [136, 142]), with approximately 13 Mha of forest removed globally each year between 2000 and 2010 [55]. For the same period, the net rate of forest area change is estimated at −5.2 million hectares per year [55]. This is lower than the rate of

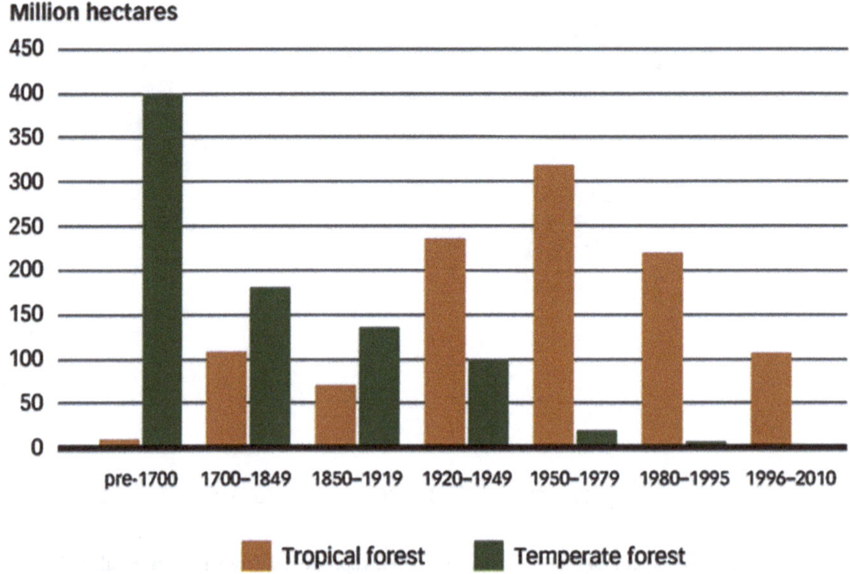


**Fig. 1.3** Estimated area of deforestation, in tropical and temperate regions, from pre-1700 to 2010; reproduced from FAO [56]

forest loss, due to extensive afforestation, particularly in China [53, 54, 239], and the natural expansion of forests onto previously managed lands.

As indicated in Fig. 1.4, the distribution of the world's forests is changing; whilst most deforestation occurs in the tropics, natural forest expansion and afforestation occurs primarily in temperate and boreal regions.

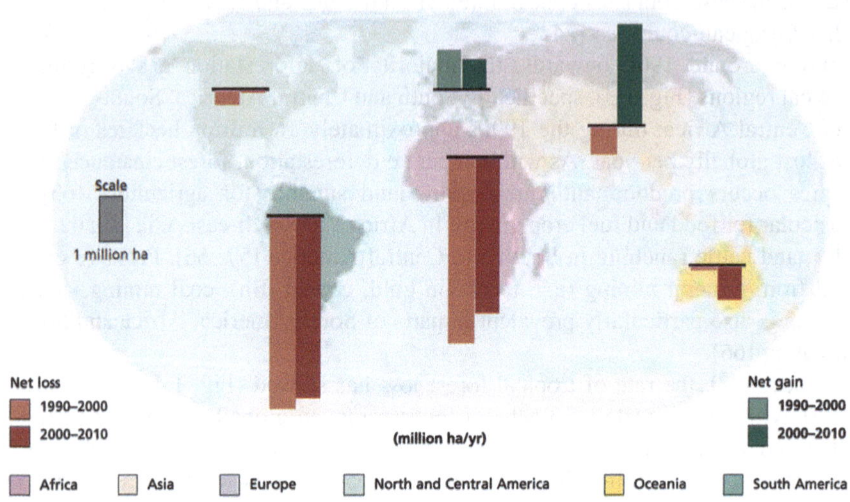


**Fig. 1.4** Annual change of forest area by region; reproduced from FAO [55]

### 1.2.2 Role of Forests in the Carbon Cycle

Plants take in carbon, in the form of carbon dioxide ($CO_2$), from the atmosphere. Approximately half of this carbon is returned to the atmosphere during respiration, whilst the other half is fixed as plant biomass during photosynthesis. The metabolic activity of ecosystems controls the atmospheric concentration of $CO_2$, with a marked seasonal cycle evident in observations [156].

The total amount of carbon stored, in terrestrial ecosystems is uncertain: the tropical forests of Latin America, Sub-Saharan Africa and Asia are estimated to contain between 247 and 553 Pg(C) [185, 203], temperate forests between 159 and 292 Pg(C) [41] and boreal forests between 395 and 559 Pg(C) [41]. Inverse modelling based on atmospheric $CO_2$ concentrations, known carbon emissions and an estimated ocean sink, suggests that, globally, the terrestrial biosphere is currently a carbon sink of approximately $\sim$1–3 Pg(C) $a^{-1}$ [31, 32, 39, 165, 185].

### 1.2.3 Biogeophysical Impacts of Forests

As well as controlling the atmospheric concentration of $CO_2$, the presence of large-scale vegetation can affect the balance of radiation in the Earth system through several biogeophysical mechanisms. The relative importance of each of these mechanisms, summarised in Fig. 1.5, is latitude dependent [25].

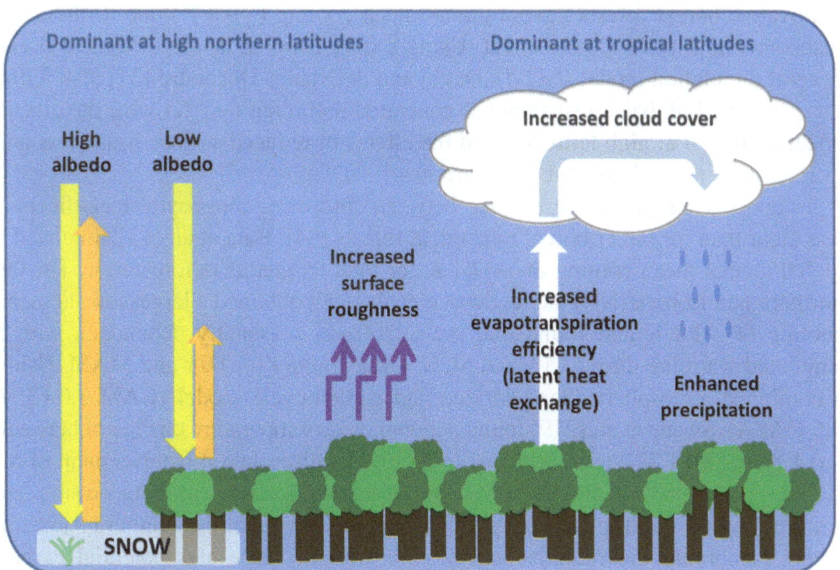


**Fig. 1.5** Summary of biogeophysical impacts of forests

Most trees have a lower albedo (typically 0.08 to 0.19 e.g. [22, 145]) than cultivated surface types such as grass or cropland (typical albedo of 0.15 to 0.26 e.g. [145]). A lower albedo means that a smaller proportion of incoming SW solar radiation is reflected by the land-surface, imposing a warming on the Earth system when compared to a higher albedo surface. At tropical latitudes, the warming induced by the lower surface albedo is more than offset by strong evaporative cooling (i.e. due to more efficient evapotranspiration by trees [211]).

At high northern latitudes (above 60°N), snow covers the ground for a substantial fraction of the year. If the snow is lying on short vegetation, it will completely cover it (Fig. 1.5) and the surface albedo will be high, e.g. 0.57 [103], or 0.75 [22]. However, if boreal trees are present, they will protrude from the snow, and the albedo will be much lower [127, 198, 221], e.g. 0.11 for snow covered spruce/poplar and 0.15 for pine [22].

Surface roughness is greater for forests (e.g. characteristic roughness length of 1 m) than for shorter vegetation (e.g. characteristic roughness length of 0.001 m), potentially increasing turbulence in the boundary layer [146]. The overall impact on surface temperature remains unclear, but a higher surface roughness could lead to increased sensible and latent heat fluxes [124].

The climatic impacts of forests are difficult to isolate from observations, so models are used to estimate the effect of removing or adding forests in a particular area [6, 13, 23, 26, 27, 34, 37, 61, 80, 124, 184, 209].

Using an integrated carbon-climate model, Bala et al. [13] found that the net climate impact of simulated global forest removal was a temperature reduction of −0.3 K. Removing tropical forest led to a global mean warming (+0.7 K) due to a reduction in evapotranspiration and high carbon storage in the tropics, whereas the removal of boreal forests gave a global mean cooling (−0.8 K) due to the dominance of the surface albedo effect. Using a fully coupled land-atmosphere-ocean general circulation model (GCM), Davin and de Noblet-Ducoudré [37] also found that simulated global forest removal generated a cooling (−1 K) with the albedo effect dominant at high latitudes, and the effects of reduced surface roughness and evapotranspiration dominant in the tropics.

In temperate regions, the balance between competing biogeophysical effects is less clear than for either boreal or tropical forests [34]. Bala et al. [13] simulated a global annual mean cooling of −0.04 K for total temperate deforestation. For the northern hemisphere (NH) alone, Snyder et al. [209] obtained a larger annual mean cooling of −1.1 K and found that the effect was seasonally dependent, with a simulated warming during JJA and SON, but a cooling for DJF and MAM. Using a combined atmospheric, land-surface and carbon-cycle model (CAM 3.0-CLM 3.5-CASA), Swann et al. [219] found a global mean temperature change of between −0.4 K and +0.1 K due to northern mid-latitude afforestation (replacement of all $C_3$ grass between 30°N and 60°N with broadleaf deciduous trees), but also simulate a northward shift of tropical precipitation belts and drying of the southern Amazon.

By combining the radiative effect of albedo decrease and potential carbon sequestration, Betts et al. [24] derived a net radiative effect for forest plantations in temperate and boreal regions (Fig. 1.6). Plantations in some regions of Canada,

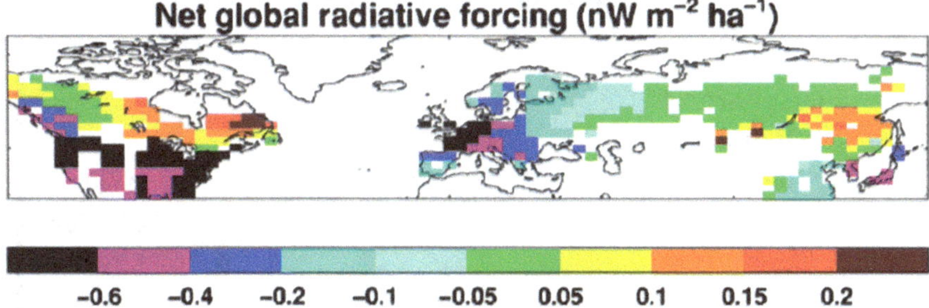


**Fig. 1.6** Estimated global net radiative forcing due to carbon sequestration and surface albedo (the value in each pixel represents the global mean net radiative effect due to afforestation of 1 ha in that pixel alone); reproduced from Betts et al. [22]

Russia and Northern China exert a simulated net positive radiative effect, whilst temperate regions of North America and Europe exert a net negative radiative effect. However, a substantial fraction of the area examined falls into the −0.05 to +0.05 nW $m^{-2}$ $ha^{-1}$ category (green in Fig. 1.6), suggesting that forests planted here could contribute a warming or cooling if there were additional climate impacts not considered by this study.

## *1.2.4 An Additional Biogeochemical Impact of Forests?*

In addition to $CO_2$, vegetation exchanges carbon with the atmosphere in the form of gas-phase biogenic volatile organic compounds (BVOCs), such as isoprene ($C_5H_8$; 2-methyl 1,3-butadiene), monoterpenes ($C_{10}H_{16}$), and sesquiterpenes ($C_{15}H_{24}$) [72, 194, 204, 243]. Once oxidised (Sect. 1.2.4.2), these compounds are able to partition into the atmospheric particle phase (Sect. 1.2.4.3). Figure 1.7 summarises how this process may affect the climate; *directly* by perturbing the path of incoming solar radiation, and *indirectly* by modifying the microphysical properties of clouds.

### 1.2.4.1 The Emission of Biogenic Volatile Organic Compounds

The production of biogenic volatile organic compounds (BVOCs) (Fig. 1.8) requires a large energy investment from plants (1–2 % of photosynthetic carbon fixation), suggesting that some form of advantage is gained from their emission; potentially by enhancing the resilience of the plant to abiotic stress (e.g. temperature, light and oxidative damage [132, 238]), inhibiting the establishment of competing plants [148], altering the climate (e.g. temperature, precipitation and cloud cover; e.g. [163, 212]), allowing below ground signalling [191], or reducing

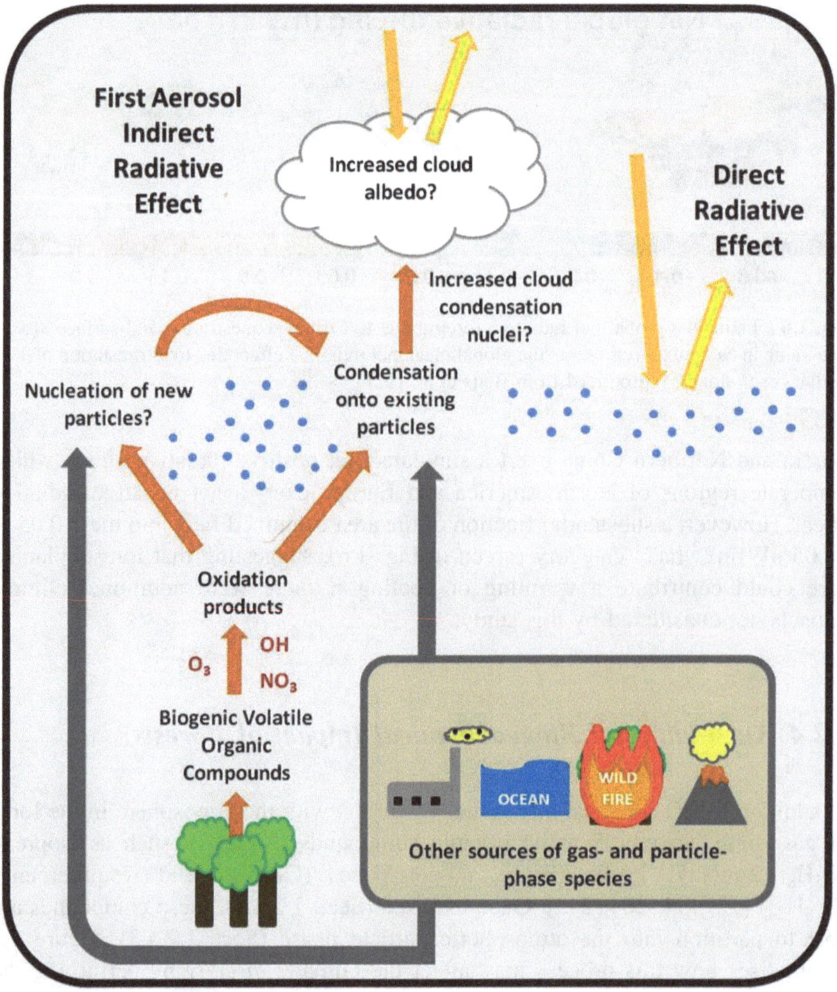

**Fig. 1.7** Summary of the additional biogeochemical impact of forests explored in this thesis

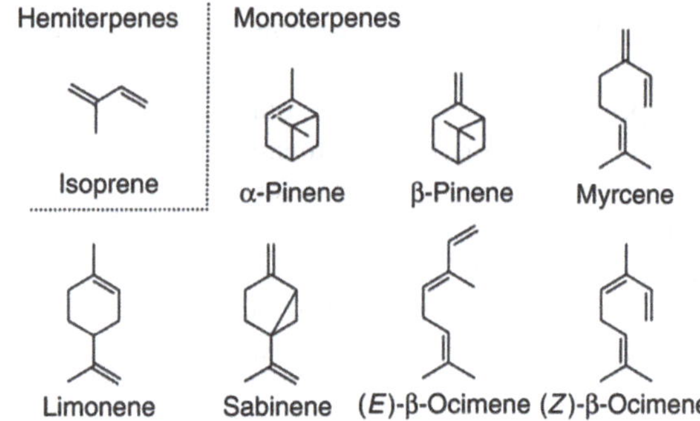


**Fig. 1.8** Chemical structures of isoprene and seven monoterpenes; reproduced from Vickers et al. [238]

insect and herbivore attack [1, 109, 161]. Insects, such as the sawfly, are known to sequester monoterpenes from their host plants and eject them in their own defence [84].

Isoprene, monoterpenes and sesquiterpenes are synthesised, de novo, during the plastidic 2-C-methyl-D-erythritol 4-phosphate (MEP) and the cytosolic mevalonic acid (MVA) biosynthetic pathways within the chloroplast [122, 128, 201]. Dimethylallyl pyrophosphate (DMAPP) is converted to isoprene by isoprene synthase [208], whereas monoterpenes and sesquiterpenes are generated by the addition of isopentyl pyrophosphate (IPP) to DMAPP, via geranyl diphosphate (GPP) and farnesyl diphosphate (FPP) respectively [138, 160]. Many other isoprenoid compounds (i.e. those composed of repeating $C_5$ units) are generated by plants, but their vapour pressure is too low to allow them to volatilise at normal biological temperatures [45].

Once produced, monoterpenes are stored in the leaf and emitted when the vapour pressure of the individual compounds allows volatilisation [38], from the adaxial side of the leaf which does not have stomatal pores [72]. Consequently, monoterpene emission levels vary exponentially with temperature, but are not thought to be strongly controlled by light availability [224]. As monoterpenes are stored in the leaf, their emission can still be detected when their synthesis may be assumed to be low e.g. due to drought [123].

Isoprene production is both light and temperature-dependent [73, 193, 223]. The observed increase in isoprene emission with increasing temperature has been attributed to the kinetics of the enzyme isoprene synthase; however isoprene synthase may become denatured at very high temperatures (>40 °C) above which isoprene emission levels fall [73, 144]. Isoprene is emitted from the abaxial side of the leaf which features stomatal pores through which the molecules diffuse [72], leaving a negligible amount stored in the leaf. Experimental studies have shown that isoprene emission is also sensitive to ambient $CO_2$ concentration [207], becoming inhibited at higher levels [202]. The lower isoprene emission level was found to accompany an increase in above ground biomass, and did not reflect a decrease in overall productivity; photosynthesis generally being enhanced under higher [$CO_2$].

The emission of BVOCs by vegetation is sensitive to disturbances, hence their hypothesised role in plant defence. Forest management i.e. logging, has been found to temporarily enhance monoterpene emission [76] and beetle infestation has also been shown to cause an overall increase in monoterpene emission [20].

In order to quantify global BVOC emissions and represent their behaviour in land-surface, climate and Earth-system models, algorithms have been developed to calculate emissions from a variety of species under different climatic conditions e.g. [4, 69, 70, 120, 181, 182, 234]; either by linking BVOC emission directly to light intensity and temperature e.g. [71], or calculating BVOC emission at the leaf level on the basis of electron availability e.g. [154, 155]. Combining these emission models with different land-cover information and climate data yields a wide range of global total BVOC emission estimates; 32–156 Tg(C) $a^{-1}$ for monoterpenes and 412–601 Tg(C) $a^{-1}$ for isoprene [5, 71].

### 1.2.4.2 BVOCs in the Atmosphere

Considerable diurnal variability is observed in the atmospheric concentration of BVOCs [78, 99]. Ambient daytime monoterpene concentrations up to 500 pptv have been observed at boreal forest locations in the NH summertime, whilst concentrations up to 8,000 pptv have been reported at night (Table 1.2). Conversely, in the Amazon rainforest, monoterpene concentrations have been observed to peak during the day at around 800 pptv [101]. The diurnal cycle in observed BVOC concentrations occurs due to controls on emissions (e.g. sensitivity to light and temperature), the height of the boundary layer and amount of vertical mixing, and the nature of reactions occurring in the atmosphere.

BVOCs are highly reactive and quickly oxidised by the hydroxyl radical (OH), ozone ($O_3$) and the nitrate radical ($NO_3$), to form more highly functionalised, but lower volatility, compounds (see Table 1.1 for atmospheric lifetimes with respect to oxidation). Atmospheric concentrations of photochemically controlled OH and $O_3$ peak during the daytime, when incoming solar radiation levels are highest [85, 206]. Additionally, the ozonolysis of BVOCs has been shown to be a source of OH in the troposphere [8, 42, 92, 172]. Whilst BVOC emissions are highest in the daytime, the lower concentrations of photochemical oxidants, and restricted vertical mixing, may allow BVOC concentrations to rise during the night [78, 99]. $NO_3$ is rapidly photolysed during the daytime, but can accumulate at night [12, 143] which, together with lower OH and $O_3$ concentrations, means that $NO_3$ oxidation dominates night-time BVOC chemistry.

These oxidation reactions proceed mainly via addition to one of the $C = C$ double bonds ($O_3$, OH and $NO_3$), or to a lesser extent, H-atom abstraction from C–H bonds ($NO_3$ and OH only) [9]. The oxidation reactions of monoterpenes have been extensively studied in the laboratory and the first-stage oxidation products are well known [10, 11, 21, 86, 93, 166, 216, 218, 249]. The OH-initiated oxidation of α-pinene proceeds mainly via OH addition to the double bond (~90 %), with a smaller contribution from hydrogen abstraction, forming pinonaldehyde, acetone, formaldehyde and other hydroxycarbonyls [77, 174]. Ozonolysis of monoterpenes follows the *Criegee* mechanism [36]; α-pinene has been shown to produce pinonic acid, norpinonic acid, pinonaldehyde and norpinonaldehyde [248, 250], via a *Criegee intermediate*. Whereas, oxidation of β-pinene by $O_3$ and OH has been shown to produce almost exclusively nopinone (6,6-dimethylbicyclo[3.1.1]heptan-2-one) and formaldehyde [7, 77, 86, 245, 248].

Laboratory oxidation of isoprene by OH and $O_3$ has been shown to yield predominantly formaldehyde, methacrolein (MTA) and methyl vinyl ketone (MVK) [9, 68, 104, 170, 171, 227]. When isoprene reacts with OH, peroxyradicals ($RO_2$) are generated which may react further with nitrogen oxide, or recombine to form peroxides. The peroxyradicals may also undergo internal reactions, yielding organic hydroperoxy radicals, eventually resulting in net OH production and leading to high OH concentrations over pristine forest regions [125, 173, 220]. Oxidation of isoprene with $NO_3$ yields mainly alkyl nitrates, but also small quantities of formaldehyde, MVK and MTA [14, 118].

**Table 1.1** Summary of the main BVOCs emitted by vegetation and their associated atmospheric lifetime with respect to oxidation (where known) by OH, $O_3$ and $NO_3$

| Biogenic volatile organic compound | | Example of known emitting genus[a] | Global emission estimate ($Tg\ a^{-1}$) | Estimated lifetime for reaction with: (assumed oxidant concentration given in brackets) | | |
|---|---|---|---|---|---|---|
| | | | | OH ($2 \times 10^6$ mol $cm^{-3}$) | $O_3$ ($7 \times 10^{11}$ mol $cm^{-3}$) | $NO_3$ ($2.5 \times 10^8$ mol $cm^{-3}$) |
| Isoprene ($C_5H_8$) | | *Eucalyptus* [72]<br>*Quercus* [222] | 535[b] | 1.4 h | 1.3 day | 1.6 h |
| Monoterpenes ($C_{10}H_{16}$) | α-pinene | *Abies* [192]<br>*Pinus* [224]<br>*Picea* [51] | 66[b] | 2.6 h | 4.6 h | 11 min |
| | β-pinene | *Pinus* [224]<br>*Picea* [51] | 19[b] | 1.8 h | 1.1 days | 27 min |
| | limonene | *Pinus* [224] | 11[b] | 49 min | 2 h | 5 min |
| | 3-carene | *Larix* [215] | 7[b] | 1.6 h | 11 h | 7 min |
| | camphene | *Picea* [51] | 4[b] | 2.6 h | 18 days | 1.7 h |
| | sabinene | *Quercus* [217] | 9[b] | 1.2 h | 4.8 h | 7 min |
| | myrcene | *Pinus* [224]<br>*Picea* [51] | 9[b] | 39 min | 50 min | 6 min |
| | *cis-/trans-*ocimene | *Pinus* [214] | 19[b] | 33 min | 44 min | 3 min |
| | α-phellandrene | *Schinus* [35] | – | 27 min | 8 min | 0.9 min |
| | β-phellandrene | *Pinus* [224]<br>*Picea* [51] | 1.5[b] | 50 min | 8.4 h | 8 min |
| | α-terpinene | *Abies* [91] | 1[c] | 23 min | 1 min | 0.5 min |
| | γ-terpinene | *Sequoia* [162] | | 47 min | 2.8 h | 2 min |
| | terpinolene | *Picea* [91] | 1.3[b] | 37 min | 13 min | 0.7 min |

(continued)

**Table 1.1** (continued)

| Biogenic volatile organic compound | | Example of known emitting genus[a] | Global emission estimate (Tg $a^{-1}$) | Estimated lifetime for reaction with: (assumed oxidant concentration given in brackets) | | |
|---|---|---|---|---|---|---|
| | | | | OH ($2 \times 10^6$ mol $cm^{-3}$) | $O_3$ ($7 \times 10^{11}$ mol $cm^{-3}$) | $NO_3$ ($2.5 \times 10^8$ mol $cm^{-3}$) |
| Sesquiterpenes ($C_{15}H_{24}$) | β-caryophyllene | *Pinus* [60] | 7[b] | 42 min | 2 min | 3 min |
| | α-farnesene | *Pinus* [60] | 7[b] | – | – | – |
| | α-copaene | *Salvia* [3] | – | 1.5 h | 2.5 h | 4 min |
| | α-humulene | *Pinus* [60] | 2[b] | 28 min | 2 min | 2 min |

[a] Common names: *Albies* fir, *Eucalyptus* eucalyptus /gum, *Larix* larch, *Quercus* oak, *Pinus* pine, *Picea* spruce, *Schinus* pepper tree, *Sequoia* redwood

[b] Reference [71]

[c] Reference [67, 69]

**Table 1.2** Summary of ambient BVOC concentration measurements

| Location | Time | | Conc. (pptv) | Reference |
|---|---|---|---|---|
| **Isoprene** | | | | |
| Pötsönvaara, Eastern Finland. Mainly *Pinus sylvestris* forest. Samples collected approximately 100 m from forest. | Jun–Aug | D | 68–346 | [79] |
| Hyytiälä, Finland. Mixed forest (*Pinus sylvestris, Picea abies, Populus tremula, Betula pubescens, Alnus incana*) | Jun–Sep | D | >140 | [123] |
| | Oct–May | | <100 | |
| **Total Monoterpenes** | | | | |
| Oslo, Norway. *Picea abies* and *Pinus sylvestris* forest. Samples collected within canopy. | June / Aug | D | 8800–70700 (pptC) | [96] |
| Jäadraås, Sweden. *Pinus sylvestris* forest | June–Aug | D | 10–500 | [98] |
| | | N | 200–8000 | |
| Hyytiälä, Finland. Mixed forest (*Pinus sylvestris, Picea abies, Populus tremula, Betula pubescens, Alnus incana*) | Oct–Apr | D | <100 | [123] |
| | Jun–Aug | | >250 | |
| | Dec–May | M | 3–96 | [78] |
| | Jun–Aug | | 129–508 | |
| | Sep–Nov | | 16–257 | |
| Manaus, Brazil. Undisturbed, mature rainforest in Central Amazonia | Sep–Dec 2010 | D | ~600–800 | [101] |
| | | N | ~200 | |
| **Total Sesquiterpenes** | | | | |
| Manaus, Brazil. Undisturbed, mature rainforest in Central Amazonia | Sep–Dec 2010 | D | 150–250 | [101] |
| | | N | 250–800 | |
| Hyytiälä, Finland. Mixed forest (*Pinus sylvestris, Picea abies, Populus tremula, Betula pubescens, Alnus incana*) | Dec–May | M | 0.1–4.2 | [78] |
| | Jun–Aug | | 2.3 | |
| | Sep–Nov | | 0.7–13 | |

*D* Indicates a day time measurement, *N* a night time measurement, and *M* a monthly mean

#### 1.2.4.3 Biogenic Secondary Organic Aerosol (SOA)

As originally suggested by Went [242], the oxidation products of BVOCs may partition into the particle phase, forming secondary organic aerosol (SOA) [102, 105, 107, 114, 158]. Semi-volatile gas-phase organic compounds may reach equilibrium with absorbing material in the particle phase [167, 168], adding to the existing organic mass in the aerosol distribution [159]. Extremely low-volatility compounds (i.e. those with saturation concentrations less than approximately $10^{-3}$ μg $m^{-3}$) may condense kinetically onto the surface of existing particles, at a rate controlled by the difference between the ambient partial pressure of the substance and its equilibrium vapour pressure over the particle surface [43, 180, 196, 254]. Very low-volatility compounds may also be formed via reactions in the particle phase, following which their re-evaporation would be inhibited [197]. The importance of these processes is discussed further in Chap. 5.

Oxidation products of isoprene, monoterpenes and sesquiterpenes have been observed in ambient aerosol [33, 100, 108, 251]. Over Scandinavia, parcels of air have been found to contain an aerosol mass that is proportional to the length of time the air has spent over forested land [228, 229]. Whilst their presence is widely observed, the exact formation mechanisms and subsequent behaviours of these semi- and low-volatility BVOC oxidation products remain unclear.

The first stage oxidation products of isoprene (e.g. MVK and MTA) are too volatile to partition into the aerosol phase, originally leading to the conclusion that isoprene oxidation does not generate SOA [166]. However, further oxidation of these first generation products has been shown to yield lower volatility 2-methyltetrols in smog chamber experiments [48], compounds which have also been observed in aerosol collected from forest locations in the Amazon [33, 240], Scandinavia [111], central Europe [97] and the USA [40]. Employing smog chamber experiments and field measurements, Lin et al. [130] suggest that methacrylic acid epoxide (MAE), formed by isoprene oxidation under low $NO_x$ conditions, via methacryloylperoxynitrate (MPAN), could act as a precursor to SOA formation. Laboratory studies suggest that SOA formation from isoprene oxidation occurs at a low mass yield, e.g. between 0.9 % and 5 % [112, 113]; depending upon conditions such as temperature, relative humidity and $NO_x$ concentration [48, 166]. However, due to the large global isoprene source ($\sim$535 Tg $a^{-1}$; [71], even a low yield would produce a substantial amount of SOA. Laboratory production of SOA from isoprene oxidation has been shown to be highly sensitive to $NO_x$ concentration; at low initial $NO_x$ concentrations (<150 ppb), SOA yields increase with increasing $NO_x$ concentration, but at high initial $NO_x$ concentration (>200 ppb) SOA yield tends to decrease with increasing $NO_x$ concentration [113].

Chamber studies of the SOA formation from monoterpene oxidation suggest higher yields than those observed for isoprene [87, 93, 153, 186]; for example, Hoffmann et al. [93] obtained mass yields of 12.5 % for α-pinene, and 30.2 % for β-pinene. For sesquiterpene oxidation, laboratory derived yields of SOA formation can be very high, for example Ng et al. [153] observed mass yields of greater than 100 % for the oxidation of longifolene.

However, extrapolating laboratory derived yields for SOA formation to the atmosphere is not straightforward. To obtain reliable yields, chamber experiments must be conducted at higher than atmospherically relevant initial BVOC concentrations in order to achieve acceptable signal-to-noise ratios [186] and overcome losses to the chamber walls [112, 113]. An alternative approach is to correlate the measured amount of aerosol mass added with an estimate of the BVOC emission

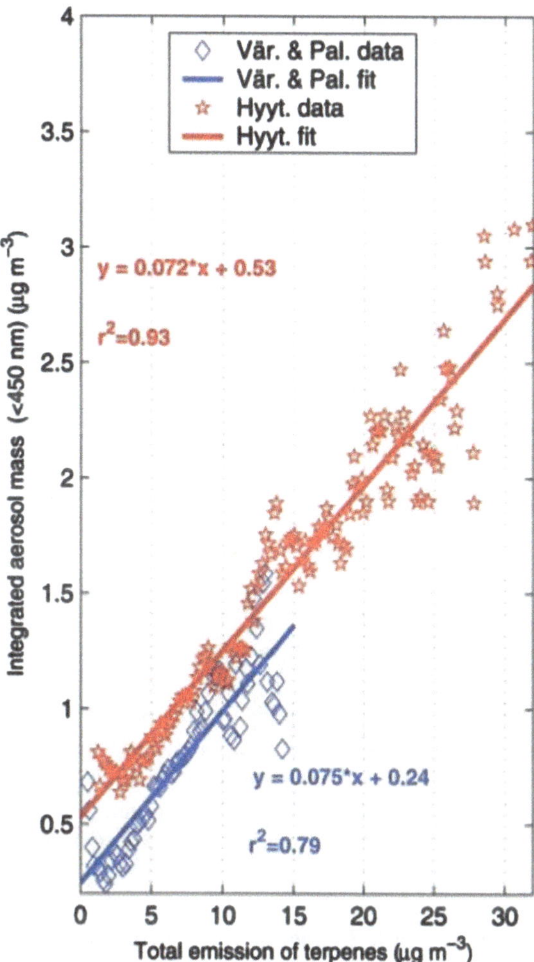


**Fig. 1.9** Average observed aerosol mass versus total estimated emission of monoterpenes at Hyytiälä (*red*), Pallas and Värriö (*blue*), in Finland; reproduced from Tunved et al. [228]

encountered during the trajectory of an air parcel. For example, Tunved et al. [228] derive an apparent mass yield for the formation of SOA from monoterpene oxidation that ranges between 5 and 10 % (Fig. 1.9). This approach, however, cannot account for the role of atmospheric sinks (of BVOCs or their oxidation products) and may therefore underestimate the yield. Equally, all increase in aerosol mass is attributed to BVOC emission, which may lead to an overestimate.

#### 1.2.4.4 Organically Mediated New Particle Formation

As well as partitioning into the existing atmospheric aerosol distribution, the oxidation products of terpenes may play a role in the initial stages of new particle formation (see Chaps. 2 and 3 for more details on new particle formation), via stabilisation of the critical nucleus [52, 115, 141, 164, 253]. In contrast, both laboratory [110] and field experiments [106] indicate that isoprene emissions may act to suppress new particle formation in the presence of high isoprene to monoterpene emission ratios, although the mechanism through which this process operates is not known.

#### 1.2.4.5 Global SOA Budget

Due to uncertainties in the distribution and strength of BVOC emission sources (Sect. 1.2.4.1), the amount of SOA produced from BVOC oxidation and the exact nature of its behaviour in the atmosphere, the global budget of SOA is poorly constrained. Estimates derived by modelling emissions and applying laboratory derived yields to BVOC oxidation suggest a global SOA production between 12 and 70 Tg(SOA) $a^{-1}$ [105], whilst top-down estimates based on satellite observations or atmospheric mass balance are an order of magnitude larger, up to 1820 Tg(SOA) $a^{-1}$ ([64, 81, 89]; assuming a conversion factor of 2 Tg(SOA)/Tg(C) to convert from literature values; Spracklen et al. [213]).

Additionally, there may be a contribution to SOA directly from anthropogenic precursors (e.g. toluene and xylene), or an anthropogenic enhancement of the biogenic source [241]. Spracklen et al. [213] derive an optimised SOA source of $140 \pm 90$ Tg(SOA) $a^{-1}$, by comparing simulated organic aerosol formation (from biogenic and anthropogenic precursors) with aerosol mass spectrometer measurements, and suggest an important role for *anthropogenically controlled* SOA. This suggestion is consistent with Heald et al. [88] who find the best agreement between simulated and observed organic aerosol mass concentrations when 100 Tg of *anthropogenically controlled* SOA is added to their simulations.

#### 1.2.4.6 Climatic Impact of Biogenic SOA

Whilst organic aerosol has been found to dominate the mass of fine aerosol at sites across the world [102, 153], its impact on the climate remains poorly constrained [133]. The presence of SOA can potentially influence the Earth's radiative balance *directly* by contributing to the absorption or scattering of radiation, and *indirectly* by altering the microphysical properties of clouds [58].

##### The Direct Radiative Effect of SOA

Particles in the atmosphere interact *directly* with incoming solar radiation; whether, and by how much, they subsequently scatter or absorb the radiation is dependent upon their size, chemical composition and optical properties [90, 175, 189]. Incoming solar radiation peaks at wavelengths between 380 nm and 750 nm, so processes that add particles of this size to the atmosphere, or aid in the growth of smaller particles to these sizes (such as the condensation of secondary organic material) will influence the path of radiation in the atmosphere.

The annual global direct radiative effect from biogenic SOA has been previously estimated at between −0.01 W $m^{-2}$ and −0.29 W $m^{-2}$ [65, 157, 190], but regional effects may be much larger, e.g. summertime mean of between −0.37 W $m^{-2}$ and −0.74 W $m^{-2}$ over boreal regions [129] and up to −1 W $m^{-2}$ over tropical forest regions [190].

The ability of a particle to scatter radiation may be described by its wavelength dependent, single scattering albedo ($\omega$), i.e. the ratio of radiation scattered to the total of radiation scattered and absorbed, and its complex refractive index ($k$). Aerosol components that are efficient at scattering radiation have a high value of $\omega$ (e.g., $\omega$ for sulphate is close to 1 at 550 nm), and efficient absorbers of radiation, such as black carbon, have a lower $\omega$.

The optical properties of SOA are not well known, owing to the range of possible compositions, and the practical limitations of measuring optical parameters under atmospherically relevant conditions. Nakayama et al. [151] found that SOA formed from the photo-oxidation of $\alpha$-pinene did not absorb UV radiation (at either 355 or 532 nm), and Lambe et al. [121] found a $\omega$ value between 0.99 and 1 for $\alpha$-pinene SOA (at 405 nm).

**The First Aerosol Indirect Radiative Effect of SOA**

The presence of particles in the atmosphere facilitates the formation of cloud droplets at much lower supersaturations (*SS*, relative humidity minus 100 %) than would be required for the homogenous nucleation of pure water droplets. Only a subset of particles, known as cloud condensation nuclei (CCN), are able to form cloud droplets under atmospherically relevant conditions.

The ability of a particle to act as a CCN, at a given *SS*, depends approximately upon the number of potential solute molecules it contains, determined by its size and chemical composition [47, 137]. At atmospherically relevant values of *SS*, across a range of compositions, CCN-active particles are typically sized between 50 and 150 nm diameter. As such, atmospheric processes that influence the amount of water-soluble material in ~50–150 nm diameter particles will affect CCN concentrations.

The presence of SOA can affect CCN number concentrations in several ways. The condensation of organic compounds is known to aid in the growth of newly formed particles to observable sizes (>3 nm) and beyond, to a CCN-active size [28, 116, 119, 179, 180, 195–197, 237, 247]. Additionally, condensing organic oxidation products may make hydrophobic particles more hydrophilic [178], and may also play a role in new particle formation (Sect. 1.2.4.4). SOA formed in the laboratory from the oxidation of isoprene and monoterpenes has been shown to exhibit CCN activity under atmospherically relevant conditions [46, 49, 50].

If the CCN concentration in the atmosphere increases, and a fixed cloud water content is assumed, the average size of cloud droplets formed will decrease, making the cloud more reflective. This is the basis of the *first aerosol indirect*, or *cloud albedo*, effect [131, 230].

Several studies suggest a large first aerosol indirect effect (AIE) from biogenic SOA over the boreal forests at high northern latitudes. Using the global aerosol microphysics model GLOMAP, Spracklen et al. [212] simulated a doubling of regional CCN concentrations as a result of monoterpene emissions, and calculated a subsequent regional AIE of between −1.8 and −6.7 W $m^{-2}$ of boreal forest. A stronger annual indirect forcing (locally between −5 and −14 W $m^{-2}$) was calculated by Kurten et al. [117] using measurements taken at a station in Finland. On a global scale, the AIE from biogenic SOA is weaker, and previous estimates range from −0.19 W $m^{-2}$ to +0.23 W $m^{-2}$ [65, 157, 190].

## 1.3 Anthropogenic Climate Change

Whilst many greenhouse gas (GHG) and aerosol species are naturally present in the Earth's atmosphere, over the past two centuries there have been significant increases in their concentrations as a result of anthropogenic activities; in particular, the large scale combustion of fossil fuels for energy, land use changes and biomass burning [58, 94, 236]. Accompanying these anthropogenic emissions has been an increase in global annual mean surface temperature of approximately 0.8 K [147, 226].

### *1.3.1 Drivers of Climate Change*

#### 1.3.1.1 Greenhouse Gases

Carbon dioxide ($CO_2$) is an unreactive gas generated during the combustion of fossil fuels and biomass, or flared directly into the atmosphere. Whilst $CO_2$ is constantly exchanged between the atmosphere, land and ocean, its lack of chemical reactivity leads to a long atmospheric residence time. Approximately 50 % of emitted $CO_2$ will be removed (i.e. partitioned amongst the ocean and land carbon sinks) from the atmosphere within the first 30 years, a further 30 % within the next few centuries, and the remainder may stay in the atmosphere for several thousands of years [39]. The global annual mean atmospheric concentration of $CO_2$ reached 394 ppmv in 2012 [156]; approximately 40 % above the pre-industrial (i.e. pre-1750) level of 280 ppm, and unprecedented during the past 650,000 years [210]. The atmospheric concentrations of other GHGs (methane, nitrous oxide, halocarbons and ozone) have also increased [58].

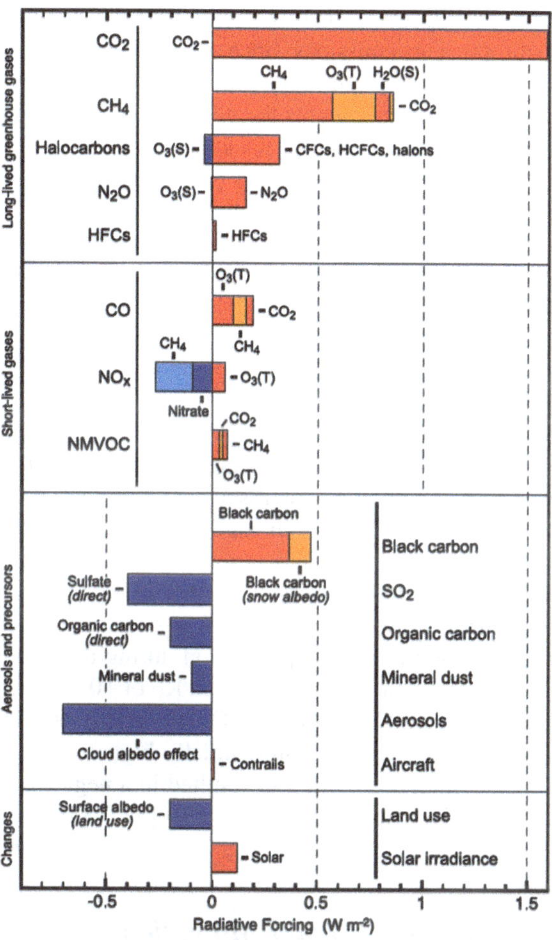


**Fig. 1.10** Components of anthropogenic radiative forcing in 2005, relative to 1750, for principal emissions; (S) and (T) represent stratospheric and tropospheric contributions respectively. Reproduced from Forster et al. [58]

Figure 1.10 details the best estimate RF (see Sect. 1.1) values attributed to the increase in atmospheric $CO_2$ concentration (+1.66 $\pm$ 0.17 W $m^{-2}$), and other long- and short-lived GHGs in 2005, since 1750, from the Fourth Assessment Report (AR4) of the Intergovernmental Panel on Climate Change [58].

#### 1.3.1.2 Aerosol Species

Whilst the estimated net anthropogenic RF from 1750 to 2005 is positive (approximately +1.6 W $m^{-2}$), there is a substantial negative RF from anthropogenically driven changes to aerosol concentrations (Fig. 1.10). The IPCC AR4 best estimate of the aerosol *direct* RF (−0.5 $\pm$ 0.4 W $m^{-2}$ [58]) represents a combination of modelled [205] and observationally [15] derived values, and is consistent with more recent model [17, 19, 150, 188, 190] and satellite-derived estimates [16,

149]. The aerosol first *indirect* RF (*cloud albedo effect* in Fig. 1.10) from anthropogenic emissions since 1750 is estimated by the IPCC at −0.7 W $m^{-2}$ (median value), with a 5–95 % range of −1.8 to −0.3 W $m^{-2}$ [58]. This wide range of values highlights the difficulty in constraining this complex parameter, as it relies upon a combined quantitative understanding of changes to global aerosol concentrations, as well as the process of cloud droplet activation under various conditions. Subsequent estimates of the first indirect RF, using more sophisticated aerosol microphysics models lie in the middle of AR4 range [17, 188, 190]; however, observationally derived values are consistently lower in absolute magnitude [18, 176, 187].

#### 1.3.1.3 Land-Use Change

As a result of the carbon stored within terrestrial ecosystems, a considerable emission of carbon can be associated with the process of land-use and land cover change (LULCC), predominantly through the decay and burning of vegetation when forest is converted to agricultural land. Over the period 1990–2009, the mean estimate of global emissions from LULCC was 1.14 ± 0.18 Pg(C) $a^{-1}$ [95], equivalent to approximately 12.5 % of annual carbon emissions from fossil fuel combustion and cement production during the 2000–2009 period [59].

Pongratz et al. [183] estimate an RF of +0.35 W $m^{-2}$ due to $CO_2$ emission from historical LULCC between AD 800 and 1992. However, since the albedo of agricultural land can be substantially higher than the forested ground it replaced, LULCC since 1750 has also resulted in a negative RF of −0.2 ± 0.2 W $m^{-2}$ due to surface albedo increase (Fig. 1.10 [58, 183]).

### *1.3.2 Climate Change Mitigation*

Using a variety of projected emission scenarios [152], and an ensemble of climate simulations, the IPCC estimated a rise in global surface temperature between 1.1 and 6.4 K by 2100 (relative to 1980–1999 mean [139]). To limit the potential future temperature increase, global anthropogenic greenhouse gas emissions must be reduced. Much climate change mitigation policy, e.g. the Copenhagen Accord [231], focuses on limiting global mean warming to 2 K above pre-industrial temperature. The aspiration to limit warming to 2 K is based on suggestions that critical thresholds may be breached and unmanageable climatic changes could occur if the global mean temperature increase exceeds 2 K; for example, widespread coral mortality, major loss of rainforests, partial deglaciation of Greenland and the West Antarctic Ice Sheet, 20–30 % of species committed to extinction, and widespread water stress in Africa and Latin America e.g. [126, 169, 199].

Due to the long atmospheric residence time of $CO_2$, cumulative emissions until any given year are of greater relevance to the $CO_2$ concentration and therefore radiative forcing, than the emissions in any specific year. Zickfeld et al. [255] found that cumulative $CO_2$ emissions from 2000 to 2500 must not exceed a median estimate of 590 Pg(C) in order to stabilise global mean temperature within 2 K of pre-industrial levels. Using a series of climate emulations for the next century, Meinshausen et al. [140] found that even if GHG emission levels in 2050 are reduced to 50 % of 2000 levels (i.e. 20 Gt($CO_2$ equivalent) $a^{-1}$), there is still up to a 49 % probability that the global mean temperature rise will exceed 2 K by 2100.

Globally, anthropogenic $CO_2$ emissions are rising each year, reaching $9.5 \pm 0.5$ Pg(C) in 2011, with 2012 emissions estimated at $9.7 \pm 0.5$ Pg(C) [177]. In 2015, the UNFCCC will review the 2 K target and potentially revise it downward to 1.5 K; however, emission trajectories that limit temperature increase to 1.5 K tend to assume that negative $CO_2$ emissions (for example through bio-energy with carbon capture and storage) will be possible by the end of the century [177, 200].

Wise et al. [246] found that the most efficient way to limit projected GHG emissions was to value carbon emissions from terrestrial land sources equally alongside those from energy and industrial systems; e.g. if only fossil fuel carbon emissions are taxed, demand for bioenergy, and therefore deforestation, will increase. Likewise, Rogelj et al. [200] concluded that "the full potential of land-based mitigation measures seems to be required in our scenarios to achieve the 2 °C target.

Due to the high heat capacity of the oceans, potential future decreases in atmospheric $CO_2$ concentration will not be accompanied by immediate temperature reductions. Using an Earth system model of intermediate complexity, Matthews and Caldeira [135] demonstrated that following an instantaneous pulse of $CO_2$, temperatures increased immediately, but did not reduce significantly for the next 500 years, despite a lack of further emissions. This suggests that to achieve a stabilisation of global temperatures it is preferable not to pass through higher concentrations of $CO_2$, on the way to $CO_2$ stabilisation, since there will be a delay in realising the subsequent temperature reductions.

#### 1.3.2.1 The Role of Forest Management in Climate Change Mitigation

A reduction in deforestation, deliberate and managed reforestation or afforestation on a large scale, could all potentially increase $CO_2$ sequestration from the atmosphere and are already being encouraged. The 1997 UN Kyoto Protocol stated that Annex 1 countries should implement policies or further elaborate existing policies such as "protection and enhancement of sinks and reservoirs of greenhouse gases not controlled by the Montreal Protocol" and "promotion of sustainable forest management practices, afforestation and reforestation". Documentation from the most recent UN Climate Change Conference in 2012 (COP-18), reiterated the sentiments of previous meetings (COP-11 onwards e.g. [232]) that efforts should be made to reduce emissions from deforestation and encourage the conservation and enhancement of carbon stocks [233]. The UN-REDD (Reducing Emissions from Deforestation and forest Degradation) programme, and the extension REDD+, aims to reduce forest loss in developing countries by introducing financial mechanisms to benefit countries that preserve the carbon stocks in their forests [2, 74]; at present, 16 developing countries are engaged in national scale REDD programmes [235]. However, there are concerns that a focus on promoting carbon storage could jeopardise the conservation of low-biomass ecosystems and human rights [29, 44].

Reducing deforestation rates by 50 % by 2050 (relative to rates observed in the 1990s), and maintaining them at that level until 2100 would avoid the direct release of approximately 50 Pg(C) [74]; equivalent to 5 years of fossil fuel carbon emissions. Using a dynamic vegetation model (LPJmL) and climatologies obtained using 5 different GCMs, Gumpenberger et al. [75] found that tropical carbon stocks in 2100 decreased by between 35 and 134 Pg(C), relative to 2012, under a continued deforestation scenario (i.e. until 50 % of forest in each grid cell remains), whereas under a forest protection scenario (i.e. forested fraction of grid cell held fixed) tropical carbon stocks increased by between 7 and 121 Pg(C).

Whilst preserved or increased forest cover would enhance $CO_2$ sequestration, forests also exert the biogeophysical impacts discussed in Sect. 1.2.3; as such the overall climatic impact of modifications to forest area will be location specific. Pongratz et al. [183] found that historical anthropogenic land-use change in temperate and boreal regions has occurred on the most productive land, thereby generating higher than average (i.e. for a particular latitude) $CO_2$ emissions. Subsequently, reforestation of these areas could induce a negative radiative effect from $CO_2$ sequestration that would outweigh the positive radiative effect of albedo increase. Arora and Montenegro [6] found that gradually replacing cropland in an Earth system model (CanESM1) with forests reduced the simulated global mean temperature for 2081–2100 by 0.45 K. The biogeophysical component of this change yielded a global mean temperature change of 0 K (with simulated warming at high latitudes balancing cooling in the tropics), and the temperature reduction occurred entirely due to increased carbon sequestration. The climate impact of changing BVOC emission levels due to land-use change, via the biogeochemical pathway described in Fig. 1.7, has not previously been assessed.

## 1.4 Aims of Thesis

The aim of this thesis is to determine the climatic significance of secondary organic aerosol formed via the oxidation of biogenic volatile organic compounds. In particular, the magnitude of the SOA effect from forests, and the implication for the role of forests in climate change mitigation will be explored.

In Chap. 2, the detailed aerosol microphysics model (GLOMAP) used in this thesis is described, with a particular emphasis on the representation of secondary organic aerosol.

In Chap. 3, this microphysics model is used to examine the role of biogenic SOA in the atmosphere and the following questions are answered:

- what impact does the presence of biogenic SOA have on particle concentrations in the present-day atmosphere?
- how sensitive is this impact to the representation of new particle formation, the amount of SOA generated from BVOC oxidation, the nature of primary carbonaceous emissions, and the presence of anthropogenic emissions?
- how does the inclusion of biogenic SOA affect agreement between observations and simulated particle concentrations?

In Chap. 4 the impact of biogenic SOA on the climate will be quantified using an offline radiative transfer model in order to establish:

- the *direct* radiative effect of the changes to particle number and size, due to biogenic SOA, determined in Chap. 3.
- the impact of these changes to particle number and size on cloud droplet number concentration.
- the *indirect* radiative effect of this change to cloud droplet number concentration.

In Chap. 5, specific assumptions concerning the volatility treatment of biogenic SOA in global models will be examined, and the following points addressed:

- how do global aerosol models differ in their representation of secondary organic aerosol?
- what are the implications of these differing representations when calculating the climate impact of biogenic SOA?

In Chap. 6, the radiative impact of biogenic SOA from particular forested regions will be quantified and compared to other forest impacts, in order to establish:

- the magnitude of the radiative effect due to forest derived biogenic SOA.
- how this radiative effect may influence the net climatic impact of forests [252].

## References

1. Amin HS et al (2013) Monoterpene emissions from bark beetle infested Engelmann spruce trees. Atmos Environ 72:130–133
2. Angelsen A, Wetrtz-Kanounnikoff S (2008) What are the key design issues for REDD and the criteria for assessing options? Moving Ahead with REDD. A. Angelsen, CIFOR, pp 11–22
3. Arey J et al (1995) Hydrocarbon emissions from natural vegetation in California's South Coast Air Basin. Atmos Environ 29(21):2977–2988
4. Arneth A et al (2007) $CO_2$ inhibition of global terrestrial isoprene emissions: potential implications for atmospheric chemistry. Geophys Res Lett 34(18):L18813

5. Arneth A et al (2008) Why are estimates of global terrestrial isoprene emissions so similar (and why is this not so for monoterpenes)? Atmos Chem Phys 8(16):4605–4620
6. Arora VK, Montenegro A (2011) Small temperature benefits provided by realistic afforestation efforts. Nat Geosci 4(8):514–518
7. Aschmann SM et al (1998) Products of the gas phase reactions of the OH radical with α- and β-pinene in the presence of NO. J Geophys Res: Atmosph 103(D19):25553–25561
8. Atkinson R (1997) Gas-phase Tropospheric chemistry of volatile organic compounds: 1. Alkanes and Alkenes. J Phys Chem Ref Data 26:215–290
9. Atkinson R, Arey J (2003) Gas-phase tropospheric chemistry of biogenic volatile organic compounds: a review. Atmos Environ 37(Supplement 2):197–219
10. Atkinson R et al (1984) Kinetics of the gas-phase reactions of nitrate radicals with a series of dialkenes, cycloalkenes, and monoterpenes at 295. ±1 K. Environ Sci Technol 18(5):370–375
11. Atkinson R et al (1990) Rate constants for the gas-phase reactions of $O_3$ with a series of monoterpenes and related compounds at 296 ± 2 K. Int J Chem Kinet 22(8):871–887
12. Atkinson R et al (1986) Estimation of night-time $N_2O_5$ concentrations from ambient $NO_2$ and $NO_3$ radical concentrations and the role of $N_2O_5$ in night-time chemistry. Atmosph Environ 20(2):331–339
13. Bala G et al (2007) Combined climate and carbon-cycle effects of large-scale deforestation. PNAS 104(16):6550–6555
14. Barnes I et al (1990) Kinetics and products of the reactions of nitrate radical with monoalkenes, dialkenes, and monoterpenes. J Phys Chem 94(6):2413–2419
15. Bellouin N et al (2005) Global estimate of aerosol direct radiative forcing from satellite measurements. Nature 438(7071):1138–1141
16. Bellouin N et al (2008) Updated estimate of aerosol direct radiative forcing from satellite observations and comparison against the Hadley Centre climate model. J Geophys Res: Atmosph 113(D10):D10205
17. Bellouin N et al (2013) Impact of the modal aerosol scheme GLOMAP-mode on aerosol forcing in the Hadley Centre Global Environmental Model. Atmos Chem Phys 13(6):3027–3044
18. Bellouin N et al (2013) Estimates of aerosol radiative forcing from the MACC re-analysis. Atmos Chem Phys 13(4):2045–2062
19. Bellouin N et al (2011) Aerosol forcing in the Climate Model Intercomparison Project (CMIP5) simulations by HadGEM2-ES and the role of ammonium nitrate. J Geophys Res 116(D20):D20206
20. Berg AR et al (2013) The impact of bark beetle infestations on monoterpene emissions and secondary organic aerosol formation in western North America. Atmos Chem Phys 13(6):3149–3161
21. Bernard F et al (2012) Thresholds of secondary organic aerosol formation by ozonolysis of monoterpenes measured in a laminar flow aerosol reactor. J Aerosol Sci 43(1):14–30
22. Betts AK, Ball JH (1997) Albedo over the boreal forest. J Geophys Res: Atmosph 102(D24):28901–28909
23. Betts RA (2000) Offset of the potential carbon sink from boreal forestation by decreases in surface albedo. Nature 408(6809):187–190
24. Betts RA et al (2007) Biogeophysical effects of land use on climate: model simulations of radiative forcing and large-scale temperature change. Agric For Meteorol 142(2–4):216–233
25. Bonan GB (2008) Forests and climate change: Forcings, feedbacks, and the climate benefits of forests. Science 320:1444–1449
26. Bonan GB et al (1992) Effects of boreal forest vegetation on global climate. Nature 359(6397):716–718
27. Bounoua L et al (2002) Effects of land cover conversion on surface climate. Clim Change 52(1–2):29–64

28. Boy M et al (2003) Nucleation events in the continental boundary layer: long-term statistical analyses of aerosol relevant characteristics. J Geophys Res 108(D21):4667
29. Brown D et al (2008) How do we achieve REDD co-benefits and avoid doing harm? In: Angelsen A (ed) Moving ahead with REDD. CIFOR. pp 107–118
30. Burgess N et al (2002) The Uluguru Mountains of eastern Tanzania: the effect of forest loss on biodiversity. Oryx 36(02):140–152
31. Canadell J, Raupach MR (2008) Managing forests for climate change mitigation. Science 320:1456–1457
32. Canadell JG et al (2007) Contributions to accelerating atmospheric $CO_2$ growth from economic activity, carbon intensity, and efficiency of natural sinks. Proc Natl Acad Sci 104(47):18866–18870
33. Claeys M et al (2004) Formation of secondary organic aerosols through photooxidation of isoprene. Science 303(5661):1173–1176
34. Claussen M et al (2001) Biogeophysical versus biogeochemical feedbacks of large-scale land cover change. Geophys Res Lett 28(6):1011–1014
35. Corchnoy SB et al (1992) Hydrocarbon emissions from twelve urban shade trees of the Los Angeles, California, Air Basin. Atmosph Environ. Part B. Urban Atmosph 26(3):339–348
36. Criegee R (1975) Mechanism of ozonolysis. Angew Chem, Int Ed Engl 14(11):745–752
37. Davin EL, de Noblet-Ducoudré N (2010) Climatic impact of global-scale deforestation: radiative versus nonradiative processes. J Clim 23(1):97–112
38. Dement WA et al (1975) Mechanism of monoterpene volatilization in Salvia mellifera. Phytochemistry 14(12):2555–2557
39. Denman KL et al (2007) Couplings between changes in the climate system and biogeochemistry. In: Solomon S, Qin D., Manninget M et al (eds.) Climate change 2007: the physical science basis. Contribution of Working Group I to the fourth assessment report of the Intergovernmental Panel on Climate Change. Cambridge University Press, Cambridge, UK and New York, USA
40. Ding X et al (2008) Spatial and seasonal trends in biogenic secondary organic aerosol tracers and water-soluble organic carbon in the Southeastern United States. Environ Sci Technol 42(14):5171–5176
41. Dixon RK et al (1994) Carbon pools and flux of global forest ecosystems. Science 263(5144):185–190
42. Donahue NM et al (1998) Direct observation of OH production from the ozonolysis of olefins. Geophys Res Lett 25(1):59–62
43. Donahue NM et al (2011) Theoretical constraints on pure vapor-pressure driven condensation of organics to ultrafine particles. Geophys Res Lett 38(16):L16801
44. Dooley K et al (2008) Cutting Corners: World Bank's forest and carbon fund fails forests and peoples, FERN / Forest Peoples Programme
45. Dudareva N et al (2006) Plant volatiles: recent advances and future perspectives. Crit Rev Plant Sci 25(5):417–440
46. Duplissy J et al (2008) Cloud forming potential of secondary organic aerosol under near atmospheric conditions. Geophys Res Lett 35(3):L03818
47. Dusek U et al (2006) Size matters more than chemistry for cloud-nucleating ability of aerosol particles. Science 312(5778):1375–1378
48. Edney EO et al (2005) Formation of 2-methyl tetrols and 2-methylglyceric acid in secondary organic aerosol from laboratory irradiated isoprene/$NO_X$/$SO_2$/air mixtures and their detection in ambient $PM_{2.5}$ samples collected in the eastern United States. Atmos Environ 39(29):5281–5289
49. Engelhart GJ et al (2008) CCN activity and droplet growth kinetics of fresh and aged monoterpene secondary organic aerosol. Atmos Chem Phys 8(14):3937–3949
50. Engelhart GJ et al (2011) Cloud condensation nuclei activity of isoprene secondary organic aerosol. J Geophys Res 116(D2):D02207
51. Evans RC et al (1985) Interspecies variation in Terpenoid emissions from Engelmann and Sitka spruce seedlings. Forest Sci 31(1):132–142

52. Fan J et al (2006) Contribution of secondary condensable organics to new particle formation: a case study in Houston, Texas. Geophys Res Lett 33(15):L15802
53. Fang J-Y et al (1998) Forest biomass of China: an estimate based on the biomass-volume relationship. Ecol Appl 8(4):1084–1091
54. Fang J et al (2001) Changes in forest biomass carbon storage in china between 1949 and 1998. Science 292(5525):2320–2322
55. FAO (2010) Global forest resources assessment 2010. FAO Forestry Paper 163. Rome, United Nations
56. FAO (2012) State of the World's Forests. Rome, United Nations
57. Fearnside PM (2005) Deforestation in Brazilian Amazonia: History, rates, and consequences. Conserv Biol 19(3):680–688
58. Forster P et al (2007) Changes in atmospheric constituents and in radiative forcing. In: Solomon S, Qin D, Manninget M et al (eds.) Climate change 2007: the physical science basis. contribution of Working Group I to the fourth assessment report of the Intergovernmental Panel on Climate Change. Cambridge University Press, Cambridge, UK and New York, USA
59. Friedlingstein P et al (2010) Update on $CO_2$ emissions. Nat Geosci 3:811–812
60. Geron CD, Arnts RR (2010) Seasonal monoterpene and sesquiterpene emissions from Pinus taeda and Pinus virginiana. Atmos Environ 44(34):4240–4251
61. Gibbard S et al (2005) Climate effects of global land cover change. Geophys Res Lett 32(23):L23705
62. Gibbs HK et al (2008) Carbon payback times for crop-based biofuel expansion in the tropics: the effects of changing yield and technology. Environ Res Lett 3(3):034001
63. Gibbs HK et al (2010) Tropical forests were the primary sources of new agricultural land in the 1980s and 1990s. In: Proceedings of the National Academy of Sciences
64. Goldstein AH, Galbally IE (2007) Known and unexplored organic constituents in the Earth's atmosphere. Environ Sci Technol 41(5):1514–1521
65. Goto D et al (2008) Importance of global aerosol modeling including secondary organic aerosol formed from monoterpene. J Geophys Res 113(D7):D07205
66. Grainger A (1993) The causes of deforestation. In: Controlling tropical deforestation, Earthscan pp 49–68
67. Griffin RJ et al (1999) Estimate of global atmospheric organic aerosol from oxidation of biogenic hydrocarbons. Geophys Res Lett 26(17):2721–2724
68. Grosjean D et al (1993) Atmospheric chemistry of isoprene and of its carbonyl products. Environ Sci Technol 27(5):830–840
69. Guenther A et al (1995) A global model of natural volatile organic compound emissions. J Geophys Res 100(D5):8873–8892
70. Guenther A et al (2006) Estimates of global terrestrial isoprene emissions using MEGAN (Model of Emissions of Gases and Aerosols from Nature). Atmos Chem Phys 6(11):3181–3210
71. Guenther AB et al (2012) The Model of Emissions of Gases and Aerosols from Nature version 2.1 (MEGAN2.1): an extended and updated framework for modeling biogenic emissions. Geosci Model Dev 5(6):1471–1492
72. Guenther AB et al (1991) Isoprene and monoterpene emission rate variability: Observations with eucalyptus and emission rate algorithm development. J Geophys Res: Atmosph 96(D6):10799–10808
73. Guenther AB et al (1993) Isoprene and monoterpene emission rate variability: model evaluations and sensitivity analyses. J Geophys Res: Atmosph 98(D7):12609–12617
74. Gullison RE et al (2007) Tropical forests and climate policy. Science 316(5827):985–986
75. Gumpenberger M et al (2010) Predicting pan-tropical climate change induced forest stock gains and losses—implications for REDD. Environ Res Lett 5(1):014013
76. Haapanala S et al (2012) Is forest management a significant source of monoterpenes into the boreal atmosphere? Biogeosciences 9(4):1291–1300

77. Hakola H et al (1994) Product formation from the gas-phase reactions of OH radicals and $O_3$ with a series of monoterpenes. J Atmos Chem 18(1):75–102
78. Hakola H et al (2012) In situ measurements of volatile organic compounds in a boreal forest. Atmos Chem Phys 12(23):11665–11678
79. Hakola H et al (2000) The ambient concentrations of biogenic hydrocarbons at a northern European, boreal site. Atmos Environ 34(29–30):4971–4982
80. Hallgren W et al (2013) Climate impacts of a large-scale biofuels expansion. Geophys Res Lett 40(8):1624–1630
81. Hallquist M et al (2009) The formation, properties and impact of secondary organic aerosol: current and emerging issues. Atmos Chem Phys 9(14):5155–5236
82. Hansen J et al (1997) Radiative forcing and climate response. J Geophys Res 102
83. Hansen MC et al (2010) Quantification of global gross forest cover loss. Proc Natl Acad Sci 107(19):8650–8655
84. Harborne JB (1988) Introduction to ecological biochemistry. Academic Press, Boston
85. Hard TM et al (1986) Diurnal cycle of tropospheric OH. Nature 322(6080):617–620
86. Hatakeyama S et al (1989) Reactions of ozone with α-pinene and β-pinene in air: yields of gaseous and particulate products. J Geophys Res: Atmosph 94(D10):13013–13024
87. Hatakeyama S et al (1991) Reactions of OH with α-pinene and β-pinene in air: estimate of global CO production from the atmospheric oxidation of terpenes. J Geophys Res: Atmosph 96(D1):947–958
88. Heald CL et al (2011) Exploring the vertical profile of atmospheric organic aerosol: comparing 17 aircraft field campaigns with a global model. Atmos Chem Phys 11(24):12673–12696
89. Heald CL et al (2010) Satellite observations cap the atmospheric organic aerosol budget. Geophys Res Lett 37(24):L24808
90. Hegg DA et al (1997) Chemical apportionment of aerosol column optical depth off the mid-Atlantic coast of the United States. J Geophys Res 102(D21):25293–25303
91. Helmig D et al (1999) Biogenic volatile organic compound emissions (BVOCs) I. Identifications from three continental sites in the U.S. Chemosphere 38(9):2163–2187
92. Herrmann F et al (2010) Hydroxyl radical (OH) yields from the ozonolysis of both double bonds for five monoterpenes. Atmos Environ 44(28):3458–3464
93. Hoffmann T et al (1997) Formation of organic aerosols from the oxidation of biogenic hydrocarbons. J Atmos Chem 26(2):189–222
94. Houghton RA (2003) Revised estimates of the annual net flux of carbon to the atmosphere from changes in land use and land management 1850–2000. Tellus B 55(2):378–390
95. Houghton RA et al (2012) Carbon emissions from land use and land-cover change. Biogeosciences 9(12):5125–5142
96. Hov Ø et al (1983) Measurement and modeling of the concentrations of terpenes in coniferous forest air. J Geophys Res: Oceans 88(C15):10679–10688
97. Ion AC et al (2005) Polar organic compounds in rural PM2.5 aerosols from K-puszta, Hungary, during a 2003 summer field campaign: sources and diel variations. Atmos Chem Phys 5(7):1805–1814
98. Janson R (1992) Monoterpene concentrations in and above a forest of Scots pine. J Atmos Chem 14(1–4):385–394
99. Janson R et al (2001) Biogenic emissions and gaseous precursors to forest aerosols. Tellus B 53(4):423–440
100. Jaoui M et al (2007) β-caryophyllinic acid: an atmospheric tracer for β-caryophyllene secondary organic aerosol. Geophys Res Lett 34(5):L05816
101. Jardine K et al (2011) Within-canopy sesquiterpene ozonolysis in Amazonia. J Geophys Res 116(D19):D19301
102. Jimenez JL et al (2009) Evolution of organic aerosols in the atmosphere. Science 326(5959):1525–1529
103. Jin Y et al (2002) How does snow impact the Albedo of vegetated land surfaces as analyzed with MODIS data? Geophys Res Lett, 29(10):12-11–12-14

104. Kamens RM et al (1982) Ozone–isoprene reactions: product formation and aerosol potential. Int J Chem Kinet 14(9):955–975
105. Kanakidou M et al (2005) Organic aerosol and global climate modelling: a review. Atmos Chem Phys 5(4):1053–1123
106. Kanawade VP et al (2011) Isoprene suppression of new particle formation in a mixed deciduous forest. Atmos Chem Phys 11(12):6013–6027
107. Kavouras IG et al (1998) Formation of atmospheric particles from organic acids produced by forests. Nature 395(6703):683–686
108. Kavouras IG et al (1999) Formation and gas/particle partitioning of monoterpenes photo-oxidation products over forests. Geophys Res Lett 26(1):55–58
109. Kessler A, Baldwin IT (2001) Defensive function of Herbivore-Induced plant volatile emissions in nature. Science 291(5511):2141–2144
110. Kiendler-Scharr A et al (2009) New particle formation in forests inhibited by isoprene emissions. Nature 461(7262):381–384
111. Kourtchev I et al (2005) Observation of 2-methyltetrols and related photo-oxidation products of isoprene in boreal forest aerosols from Hyytiälä, Finland. Atmos Chem Phys 5(10):2761–2770
112. Kroll JH et al (2005) Secondary organic aerosol formation from isoprene photooxidation under high-NOx conditions. Geophys Res Lett 32(18):L18808
113. Kroll JH et al (2006) Secondary organic aerosol formation from isoprene photooxidation. Environ Sci Technol 40(6):1869–1877
114. Kroll JH, Seinfeld JH (2008) Chemistry of secondary organic aerosol: formation and evolution of low-volatility organics in the atmosphere. Atmos Environ 42(16):3593–3624
115. Kulmala M et al (2013) Direct observations of atmospheric aerosol nucleation. Science 339(6122):943–946
116. Kulmala M et al (2004) Formation and growth rates of ultrafine atmospheric particles: a review of observations. J Aerosol Sci 35(2):143–176
117. Kurten T et al (2003) Estimation of different forest-related contributions to the radiative balance using observations in southern Finland. Boreal Environ Res 8:275–285
118. Kwok ESC et al (1996) Product formation from the reaction of the $NO_3$ radical with isoprene and rate constants for the reactions of methacrolein and methyl vinyl ketone with the NO3 radical. Int J Chem Kinet 28(12):925–934
119. Laaksonen A et al (2008) The role of VOC oxidation products in continental new particle formation. Atmos Chem Phys 8(10):2657–2665
120. Lamb B et al (1987) A national inventory of biogenic hydrocarbon emissions. Atmosph Environ (1967) 21(8):1695–1705
121. Lambe AT et al (2013) Relationship between oxidation level and optical properties of secondary organic aerosol. Environ Sci Technol 47(12):6349–6357
122. Lange BM et al (2000) Isoprenoid biosynthesis: the evolution of two ancient and distinct pathways across genomes. Proc Natl Acad Sci 97(24):13172–13177
123. Lappalainen HK et al (2009) Day-time concentrations of biogenic volatile organic compounds in a boreal forest canopy and their relation to environmental and biological factors. Atmos Chem Phys 9(15):5447–5459
124. Lean J, Warrilow DA (1989) Simulation of the regional climatic impact of Amazon deforestation. Nature 342:411–413
125. Lelieveld J et al (2008) Atmospheric oxidation capacity sustained by a tropical forest. Nature 452(7188):737–740
126. Lenton TM et al (2008) Tipping elements in the Earth's climate system. Proc Natl Acad Sci 105(6):1786–1793
127. Leonard RE, Eschner AR (1968) Albedo of intercepted snow. Water Resour Res 4(5):931–935
128. Lichtenthaler HK et al (1997) Biosynthesis of isoprenoids in higher plant chloroplasts proceeds via a mevalonate-independent pathway. FEBS Lett 400(3):271–274

129. Lihavainen H et al (2009) Observational signature of the direct radiative effect by natural boreal forest aerosols and its relation to the corresponding first indirect effect. J Geophys Res 114:D20206
130. Lin Y-H et al (2013) Epoxide as a precursor to secondary organic aerosol formation from isoprene photooxidation in the presence of nitrogen oxides. Proc Natl Acad Sci 110(17):6718–6723
131. Lohmann U, Feichter J (2005) Global indirect aerosol effects: a review. Atmos Chem Phys 5(3):715–737
132. Loreto F, Velikova V (2001) Isoprene produced by leaves protects the photosynthetic apparatus against ozone Damage, quenches ozone products, and reduces lipid peroxidation of cellular membranes. Plant Physiol 127(4):1781–1787
133. Mahowald N et al (2011) Aerosol impacts on climate and biogeochemistry. Annu Rev Environ Resour 36(1):45–74
134. Malhi Y et al (2013) African rainforests: past, present and future. Philos Trans Royal Soc B: Biolog Sci 368(1625)
135. Matthews HD, Caldeira K (2008) Stabilizing climate requires near-zero emissions. Geophys Res Lett 35(4):L04705
136. Mayaux P et al (2013) State and evolution of the African rainforests between 1990 and 2010. Philos Trans Royal Soc B: Biolog Sci 368(1625)
137. McFiggans G et al (2006) The effect of physical and chemical aerosol properties on warm cloud droplet activation. Atmos Chem Phys 6(9):2593–2649
138. McGarvey DJ, Croteau R (1995) Terpenoid metabolism. The Plant Cell Online 7(7):1015–1026
139. Meehl GA et al (2007) Global climate projections. In: Solomon S, Qin D, Manninget M et al (eds) Climate change 2007: the physical science basis. Contribution of Working Group I to the fourth assessment report of the Intergovernmental Panel on Climate Change. Cambridge University Press, Cambridge
140. Meinshausen M et al (2009) Greenhouse-gas emission targets for limiting global warming to 2 °C. Nature 458(7242):1158–1162
141. Metzger A et al (2010) Evidence for the role of organics in aerosol particle formation under atmospheric conditions. Proc Natl Acad Sci 107(15):6646–6651
142. Miettinen J et al (2011) Deforestation rates in insular Southeast Asia between 2000 and 2010. Glob Change Biol 17(7):2261–2270
143. Mihelcic D et al (1993) Simultaneous measurements of peroxy and nitrate radicals at Schauinsland. J Atmos Chem 16(4):313–335
144. Monson RK et al (1992) Relationships among isoprene emission rate, photosynthesis, and isoprene synthase activity as influenced by temperature. Plant Physiol 98:1175–1180
145. Monteith JL, Unsworth MH (2008a) Microclimatology of Radiation (i). In: Principles of Environmental Physics. Academic Press, Boston, pp 86–99
146. Monteith JL, Unsworth MH (2008b) Micrometeorology (i). In: Principles of Environmental Physics, Academic Press, Boston pp 300–334
147. Morice CP et al (2012) Quantifying uncertainties in global and regional temperature change using an ensemble of observational estimates: the HadCRUT4 data set. J Geophys Res: Atmos 117(D8):D08101
148. Muller CH (1966) The role of chemical inhibition (Allelopathy) in vegetational composition. Bull Torrey Bot Club 93(5):332–351
149. Myhre G (2009) Consistency between satellite-derived and modeled estimates of the direct aerosol effect. Science 325(5937):187–190
150. Myhre G et al (2013) Radiative forcing of the direct aerosol effect from AeroCom Phase II simulations. Atmos Chem Phys 13(4):1853–1877
151. Nakayama T et al (2010) Laboratory studies on optical properties of secondary organic aerosols generated during the photooxidation of toluene and the ozonolysis of α-pinene. J Geophys Res: Atmos 115(D24):D24204

152. Nakicenovic N et al (2000) Special report on emission scenarios. In: Nakicenovic N, Swart R (eds)
153. Ng NL et al (2007) Effect of NOx level on secondary organic aerosol (SOA) formation from the photooxidation of terpenes. Atmos Chem Phys 7(19):5159–5174
154. Niinemets Ü et al (2002) A model coupling foliar monoterpene emissions to leaf photosynthetic characteristics in Mediterranean evergreen Quercus species. New Phytol 153(2):257–275
155. Niinemets Ü et al (1999) A model of isoprene emission based on energetic requirements for isoprene synthesis and leaf photosynthetic properties for Liquidambar and Quercus. Plant, Cell Environ 22(11):1319–1335
156. NOAA (2013) Trends in atmospheric carbon dioxide. Retrieved 17/07/2013, from http://www.esrl.noaa.gov/gmd/ccgg/trends/
157. O'Donnell D et al (2011) Estimating the direct and indirect effects of secondary organic aerosols using ECHAM5-HAM. Atmos Chem Phys 11(16):8635–8659
158. O'Dowd CD et al (2002) Aerosol formation: atmospheric particles from organic vapours. Nature 416(6880):497–498
159. Odum JR et al (1996) Gas/Particle partitioning and secondary organic aerosol yields. Environ Sci Technol 30(8):2580–2585
160. Ogura K, Koyama T (1998) Enzymatic aspects of isoprenoid chain elongation. Chem Rev 98(4):1263–1276
161. Oh KH et al (1967) Effect of various essential oils isolated from Douglas fir needles upon sheep and deer rumen microbial activity. Appl Microbiol 15:777–784
162. Okamoto RA et al (1981) Volatile terpenes in Sequoia sempervirens foliage. Changes in composition during maturation. J Agric Food Chem 29(2):324–326
163. Paasonen P et al (2013) Warming-induced increase in aerosol number concentration likely to moderate climate change. Nat Geosci 6(6):438–442
164. Paasonen P et al (2010) On the roles of sulphuric acid and low-volatility organic vapours in the initial steps of atmospheric new particle formation. Atmos Chem Phys 10(22):11223–11242
165. Pan Y et al (2011) A large and persistent carbon sink in the world's forests. Science 333(6045):988–993
166. Pandis SN et al (1991) Aerosol formation in the photooxidation of isoprene and β-pinene. Atmos Environ Part A. General Topics 25(5–6):997–1008
167. Pankow JF (1994) An absorption model of gas/particle partitioning of organic compounds in the atmosphere. Atmos Environ 28(2):185–188
168. Pankow JF (1994) An absorption model of the gas/aerosol partitioning involved in the formation of secondary organic aerosol. Atmos Environ 28(2):189–193
169. Parry ML et al (2007) Technical summary. In: Parry ML, Canziani OF, Palutikof JP, Linden VD, Hanson CE (eds) Climate change 2007: impacts, adaptation and vulnerability. Contribution of Working Group II to the fourth assessment report of the Intergovernmental Panel on Climate Change. Cambridge University Press, Cambridge
170. Paulson SE et al (1992) Atmospheric photooxidation of isoprene part I: the hydroxyl radical and ground state atomic oxygen reactions. Int J Chem Kinet 24(1):79–101
171. Paulson SE et al (1992) Atmospheric photooxidation of isoprene part II: the ozone-isoprene reaction. Int J Chem Kinet 24(1):103–125
172. Paulson SE et al (1997) Evidence for formation of OH radicals from the reaction of $O_3$ with alkenes in the gas phase. Geophys Res Lett 24(24):3193–3196
173. Peeters J et al (2009) HOx radical regeneration in the oxidation of isoprene. Phys Chem Cheml Phys 11(28):5935–5939
174. Peeters J et al (2001) The detailed mechanism of the OH-initiated atmospheric oxidation of α-pinene: a theoretical study. Phys Chem Chem Phys 3(24):5489–5504
175. Penner JE et al (2001) Aerosols, their direct and indirect effects. In: Houghton JT, Ding Y, Griggset DJ et al (eds) Climate change 2001: the physical science basis. Contribution of

Working Group I to the third assessment report of the Intergovernmental Panel on Climate Change. Cambridge University Press, Cambridge
176. Penner JE et al (2011) Satellite methods underestimate indirect climate forcing by aerosols. Proc Natl Acad Sci 108(33):13404–13408
177. Peters GP et al (2013) The challenge to keep global warming below 2 degrees C. Nature Clim Change 3(1):4–6
178. Petters MD et al (2006) Chemical aging and the hydrophobic-to-hydrophilic conversion of carbonaceous aerosol. Geophys Res Lett 33(24):L24806
179. Pierce JR et al (2012) Nucleation and condensational growth to CCN sizes during a sustained pristine biogenic SOA event in a forested mountain valley. Atmos Chem Phys 12(7):3147–3163
180. Pierce JR et al (2011) Quantification of the volatility of secondary organic compounds in ultrafine particles during nucleation events. Atmos Chem Phys 11(17):9019–9036
181. Pierce T et al (1998) Influence of increased isoprene emissions on regional ozone modeling. J Geophys Res: Atmos 103(D19):25611–25629
182. Pierce TE, Waldruff PS (1991) PC-BEIS: a personal computer version of the biogenic emissions inventory system. J Air Waste Manag Assoc 41(7):937–941
183. Pongratz J et al (2011) Past land use decisions have increased mitigation potential of reforestation. Geophys Res Lett 38(15):L15701
184. Pongratz J et al (2010) Biogeophysical versus biogeochemical climate response to historical anthropogenic land cover change. Geophys Res Lett 37(8):L08702
185. Prentice IC et al (2001) The carbon cycle and atmospheric carbon dioxide. In: Houghton JT, Ding Y, Griggset DJ et al (eds.) Climate change 2001: the physical science basis. Contribution of Working Group I to the third assessment report of the Intergovernmental Panel on Climate Change. Cambridge University Press, Cambridge
186. Presto AA et al (2005) Secondary organic aerosol production from terpene ozonolysis 2 effect of nox concentration. Environ Sci Technol 39(18):7046–7054
187. Quaas J et al (2008) Satellite-based estimate of the direct and indirect aerosol climate forcing. J Geophys Res: Atmosph 113(D5):D05204
188. Quaas J et al (2009) Aerosol indirect effects—general circulation model intercomparison and evaluation with satellite data. Atmos Chem Phys 9(22):8697–8717
189. Ramaswamy V et al (2001) Radiative forcing. In: Houghton JT, Ding Y, Griggset DJ al (eds) Climate change 2001: the physical science basis. Contribution of Working Group I to the third assessment report of the Intergovernmental Panel on Climate Change. Cambridge University Press, Cambridge
190. Rap A et al (2013) Natural aerosol direct and indirect radiative effects. Geophys Res Lett 40:3297–3301
191. Rasmann S et al (2005) Recruitment of entomopathogenic nematodes by insect-damaged maize roots. Nature 434(7034):732–737
192. Rasmussen RA (1972) What do the hydrocarbons from trees contribute to air pollution? J Air Pollution Control Assoc 22(7):537–543
193. Rasmussen RA, Jones CA (1973) Emission isoprene from leaf discs of hamamelis. Phytochemistry 12(1):15–19
194. Rasmussen RA, Went FW (1965) Volatile organic material of plant origin in the atmosphere. Proc Natl Acad Sci 53:215–220
195. Riccobono F et al (2012) Contribution of sulfuric acid and oxidized organic compounds to particle formation and growth. Atmos Chem Phys 12(20):9427–9439
196. Riipinen I et al (2011) Organic condensation: a vital link connecting aerosol formation to cloud condensation nuclei (CCN) concentrations. Atmos Chem Phys 11(8):3865–3878
197. Riipinen I et al (2012) The contribution of organics to atmospheric nanoparticle growth. Nature Geosci 5(7):453–458
198. Robinson DA, Kukla G (1985) Maximum surface albedo of seasonally snow-covered lands in the northern hemisphere. J Climate Appl Meteorol 24(5):402–411
199. Rockstrom J et al (2009) A safe operating space for humanity. Nature 461(7263):472–475

200. Rogelj J et al (2013) 2020 emissions levels required to limit warming to below 2 degrees C. Nature Clim Change 3(4):405–412
201. Rohmer M (1999) The discovery of a mevalonate-independent pathway for isoprenoid biosynthesis in bacteria, algae and higher plants. Nat Prod Rep 16(5):565–574
202. Rosenstiel TN et al (2003) Increased $CO_2$ uncouples growth from isoprene emission in an agriforest ecosystem. Nature 421(6920):256–259
203. Saatchi SS et al (2011) Benchmark map of forest carbon stocks in tropical regions across three continents. Proc Natl Acad Sci 108(24):9899–9904
204. Sanadze GA, Kursanov AL (1966) On certain conditions of the evolution of the diene $C_5H_8$ from poplar leaves. Sov Plant Physiol 13:184–189
205. Schulz M et al (2006) Radiative forcing by aerosols as derived from the AeroCom present-day and pre-industrial simulations. Atmos Chem Phys 6(12):5225–5246
206. Seinfeld JH, Pandis SN (2006) Chemistry of the Troposphere. In: atmospheric chemistry and physics: from air pollution to climate change. Wiley, New York pp 204–279
207. Sharkey TD et al (1991) High carbon dioxide and sun/shade effects on isoprene emission from oak and aspen tree leaves. Plant Cell Environ 14(3):333–338
208. Silver GM, Fall R (1991) Enzymatic Synthesis of Isoprene from Dimethylallyl Diphosphate in Aspen leaf extracts. Plant Physiol 97(4):1588–1591
209. Snyder PK et al (2004) Evaluating the influence of different vegetation biomes on the global climate. Clim Dyn 23:279–302
210. Solomon S et al (2007) Technical Summary. In: Solomon S, Qin D, Manninget M et al (eds) Climate change 2007: the physical science basis. Contribution of Working Group I to the fourth assessment report of the Intergovernmental Panel on Climate Change. Cambridge University Press, Cambridge
211. Spracklen DV et al (2012) Observations of increased tropical rainfall preceded by air passage over forests. Nature 489(7415):282–285
212. Spracklen DV et al (2008) Boreal forests, aerosols and the impacts on clouds and climate. Philos Trans Royal Soc A 366:4613–4626
213. Spracklen DV et al (2011) Aerosol mass spectrometer constraint on the global secondary organic aerosol budget. Atmos Chem Phys 11(23):12109–12136
214. Staudt M et al (1997) Seasonal and diurnal patterns of monoterpene emissions from Pinus pinea (L.) under field conditions. Atmos Environ 31(1):145–156
215. Steinbrecher R et al (1993) Terpenoid emissions from common oak (Quercus robur L.) and Norway Spruce (Picea abies L. Karst.). Air pollution research report 47: Joint Workshop CEC/BIATEX of EUROTRAC: 251–259
216. Stewart DJ et al (2013) The kinetics of the gas-phase reactions of selected monoterpenes and cyclo-alkenes with ozone and the $NO_3$ radical. Atmos Environ 70:227–235
217. Street RA et al (1997) Effect of habitat and age on variations in volatile organic compound (VOC) emissions from Quercus ilex and Pinus pinea. Atmos Environ 31(1):9–100
218. Sugarman N, Daugherty PM (1956) Oxidation of alpha-pinene. Ind Eng Chem 48(10):1831–1835
219. Swann ALS et al (2012) Mid-latitude afforestation shifts general circulation and tropical precipitation. Proc Natl Acad Sci 109(3):712–716
220. Taraborrelli D et al (2012) Hydroxyl radical buffered by isoprene oxidation over tropical forests. Nature Geosci 5(3):190–193
221. Thomas G, Rowntree PR (1992) The boreal forests and climate. Q J Royal Meteorol Soc 118(505):469–497
222. Tingey D et al (1981) Effects of environmental conditions on isoprene emission from live oak. Planta 152(6):565–570
223. Tingey DT et al (1979) The influence of light and temperature on isoprene emission rates from live oak. Physiol Plant 47(2):112–118
224. Tingey DT et al (1980) Influence of light and temperature on monoterpene emission rates from Slash pine. Plant Physiol 65(5):797–801

225. Trenberth KE et al (2009) Earth's global energy budget. Bull Am Meteorol Soc 90(3):311–323
226. Trenberth KE et al (2007) Observations: surface and atmospheric climate change. In: Solomon S, Qin D, Manninget M, et al (eds) Climate change 2007: the physical science basis. Contribution of Working Group I to the fourth assessment report of the Intergovernmental Panel on Climate Change. Cambridge University Press, Cambridge
227. Tuazon EC, Atkinson R (1990) A product study of the gas-phase reaction of Isoprene with the OH radical in the presence of NOx. Int J Chem Kinet 22(12):1221–1236
228. Tunved P et al (2006) High natural aerosol loading over boreal forests. Science 312(5771):261–263
229. Tunved P et al (2008) The natural aerosol over Northern Europe and its relation to anthropogenic emissions—implications of important climate feedbacks. Tellus B 60(4): 473–484
230. Twomey S (1977) Influence of pollution on shortwave albedo clouds. J Atmos Sci 34(7):1149–1152
231. UNFCCC (2009) Copenhagen accord
232. UNFCCC (2010) Report of the conference of the parties on its sixteenth session. United Nations
233. UNFCCC (2012) Report of the Conference of the Parties on its eighteenth session. United Nations
234. Unger N et al (2013) Photosynthesis-dependent isoprene emission from leaf to planet in a global carbon–chemistry–climate model. Atmos Chem Phys Discuss 13(7):17717–17791
235. United Nations (2013) Retrieved 12/07/2013, from http://www.un-redd.org
236. van der Werf GR et al (2004) Continental-scale partitioning of fire emissions during the 1997 to 2001 El Niño/La Niña period. Science 303(5654):73–76
237. Verheggen B et al (2007) α-pinene oxidation in the presence of seed aerosol: estimates of nucleation rates, growth rates, and yield. Environ Sci Technol 41(17):6046–6051
238. Vickers CE et al (2009) A unified mechanism of action for volatile isoprenoids in plant abiotic stress. Nat Chem Biol 5(5):283–291
239. Wang S et al (2007) Carbon sinks and sources in China's forests during 1901–2001. J Environ Manage 85(3):524–537
240. Wang W et al (2005) Characterization of oxygenated derivatives of isoprene related to 2-methyltetrols in Amazonian aerosols using trimethylsilylation and gas chromatography/ion trap mass spectrometry. Rapid Commun Mass Spectrom 19(10):1343–1351
241. Weber RJ et al (2007) A study of secondary organic aerosol formation in the anthropogenic-influenced southeastern United States. J Geophys Res 112(D13):D13302
242. Went FW (1960) Blue hazes in the atmosphere. Nature 187:641–643
243. Went FW (1960) Organic matter in the atmosphere, and its possible relation to petroleum formation. Proc Natl Acad Sci 46:212–221
244. Williams M (2003) Deforesting the earth. The University of Chigaco Press, Chigaco
245. Winterhalter R et al (2000) Products and mechanism of the gas phase reaction of ozone with β-pinene. J Atmos Chem 35(2):165–197
246. Wise M et al (2009) Implications of limiting $CO_2$ concentrations for land use and energy. Science 324(5931):1183–1186
247. Yli-Juuti T et al (2011) Growth rates of nucleation mode particles in Hyytiälä during 2003–2009: variation with particle size, season, data analysis method and ambient conditions. Atmos Chem Phys 11(24):12865–12886
248. Yokouchi Y, Ambe, Y (1985) Aerosols formed from the chemical reaction of monoterpenes and ozone. Atmos Environ (1967), 19(8):1271–1276
249. Yu J et al (1999) Gas-phase ozone oxidation of monoterpenes: gaseous and particulate products. J Atmos Chem 34(2):207–258
250. Yu J et al (1998) Identification of products containing −COOH, −OH, and −CO in atmospheric oxidation of hydrocarbons. Environ Sci Technol 32(16):2357–2370

251. Yu J et al (1999) Observation of gaseous and particulate products of monoterpene oxidation in forest atmospheres. Geophys Res Lett 26(8):1145–1148
252. Zhang Q et al (2007) Ubiquity and dominance of oxygenated species in organic aerosols in anthropogenically-influenced Northern Hemisphere midlatitudes. Geophys Res Lett 34(13):L13801
253. Zhang R et al (2009) Formation of nanoparticles of blue haze enhanced by anthropogenic pollution. Proc Natl Acad Sci 106(42):17650–17654
254. Zhang X et al (2012) Diffusion-limited versus quasi-equilibrium aerosol growth. Aerosol Sci Technol 46(8):874–885
255. Zickfeld K et al (2009) Setting cumulative emissions targets to reduce the risk of dangerous climate change. Proc Natl Acad Sci USA 106(38):16129–16134

# Chapter 2
# Model Description

This chapter describes the global aerosol microphysics model (GLOMAP-mode) used in Chaps. 3–6. An offline radiative transfer model and land-surface model are introduced and described in Chaps. 4 and 6, respectively.

## 2.1 Introduction to GLOMAP

The GLObal Model of Aerosol Processes (GLOMAP) was developed at the University of Leeds [63, 73, 79, 80] and is an extension to the TOMCAT three-dimensional, Eulerian, chemical transport model [12]. TOMCAT-GLOMAP has a horizontal resolution of 2.8° × 2.8° (T42) with 31 $\sigma$-pressure levels (i.e., terrain following at the surface), extending to 10 hPa. GLOMAP is a global aerosol microphysics model, calculating the size, number concentration and chemical composition of aerosol. It includes representations of nucleation, particle growth via coagulation, condensation and cloud processing, and wet/dry deposition; these processes are summarised in Fig. 2.1 and will be described in detail in the following chapter.

### *2.1.1 Representation of the Aerosol Size Distribution*

The work described in this thesis uses the modal version of GLOMAP [53], in which information about aerosol component masses and number concentrations (i.e., two moment) is carried in five log-normal size modes, based on the "pseudo modal" M7/HAM approach [81, 87]. This modal approach, depicted in Fig. 2.2, contrasts with the original sectional version of GLOMAP, in which the aerosol distribution is represented by 20 size sections, or bins (e.g. [17, 40, 55, 64, 75, 79]).



C.E. Scott, *The Biogeochemical Impacts of Forests and the Implications for Climate Change Mitigation*, Springer Theses, DOI 10.1007/978-3-319-07851-9_2

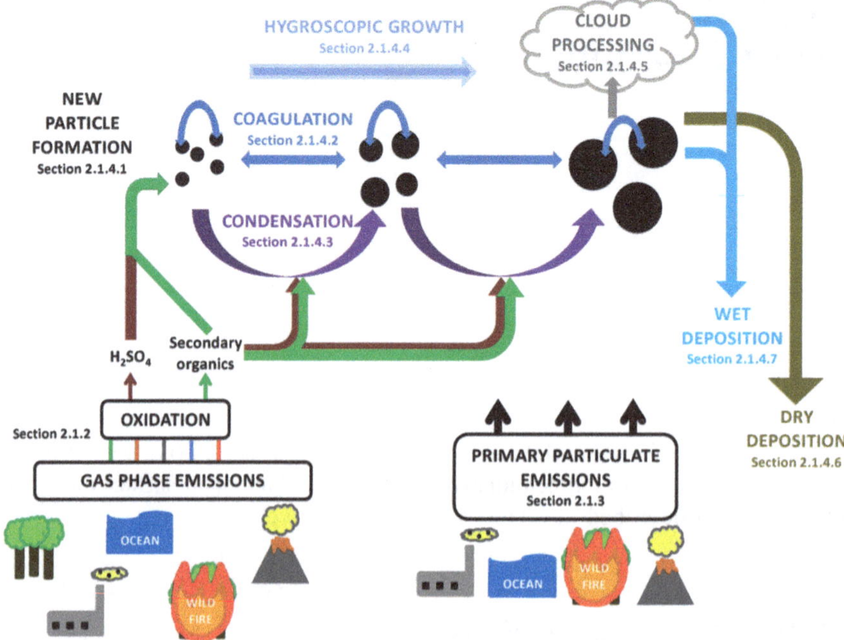


Fig. 2.1 Summary of main processes included in the GLOMAP model

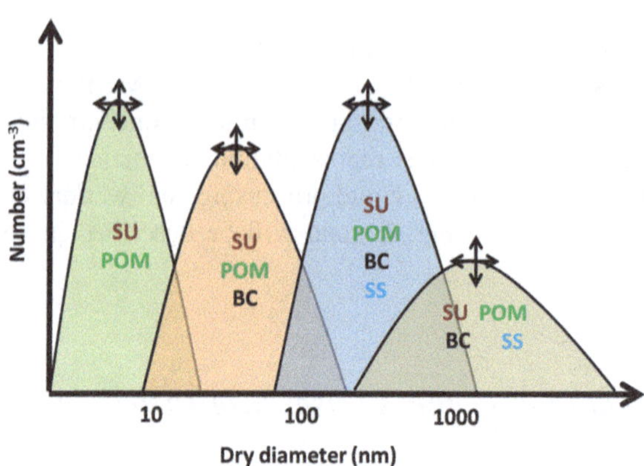


**Fig. 2.2** Schematic of the aerosol number size distribution, as represented by GLOMAP-mode. Components are classified as: sulphate (*SU*), particulate organic matter (*POM*), black carbon (*BC*) and sea-salt (*SS*)

The modal scheme is designed to allow longer integrations with greater computational efficiency and has been shown to compare well with the sectional version and observations [10, 52, 53, 67, 90].

**Table 2.1** Summary of GLOMAP-mode configuration; $D_g$ is the geometric mean diameter and $\sigma_g$ is the geometric standard deviation of each mode

| Mode | Size range | Components | Treated as soluble? | Geometric standard deviation, $\sigma_g$ |
|---|---|---|---|---|
| Nucleation | $D_g < 10$ nm | SU, POM | Yes | 1.59 |
| Aitken | $10 < D_g < 100$ nm | SU, BC, POM | Yes | 1.59 |
| | | BC, POM | No | 1.59 |
| Accumulation | 100 nm $< D_g <$ 1 μm | SU, BC, POM, SS | Yes | 1.4 |
| Coarse | $D_g >$ 1 μm | SU, BC, POM, SS | Yes | 2.0 |

Based on Mann et al. [53], as modified by Mann et al. [52]

In the configuration used for this work, material in the particle phase is classified as one of four components: sulphate (SU), black carbon (BC), particulate organic matter (POM) and sea-salt (SS), with each component allowed into the modes specified in Table 2.1. Four of the size modes are treated as hydrophilic (nucleation, Aitken, accumulation and coarse), with an additional non-hydrophilic Aitken mode.

Each mode has a fixed geometric mean standard deviation ($\sigma_g$) and contains particles with a range of geometric mean diameters ($D_g$), as described in Table 2.1. $D_g$ for each mode $i$ is calculated as in Eq. 2.1, where $V_{dry_i}$ is the total dry volume over all components $j$ in that mode.

$$D_{g_i} = \left(\frac{6V_{dry_i}}{\pi \exp\left(4.5\log^2\sigma_{g,i}\right)}\right)^{\frac{1}{3}} \tag{2.1}$$

$V_{dry_i}$ is calculated as in Eq. 2.2, according to the number of molecules per particle of component ($m_{ij}$), Avogadro's constant ($N_a$), and the density ($\rho_j$) and molar mass ($M_j$) of each component, given in Table 2.2.

$$V_{dry_i} = \sum_j \left(\frac{m_{ij}M_j}{N_a\rho_j}\right) \tag{2.2}$$

$D_g$ for each mode varies with time as the aerosol size distribution evolves; when $D_g$ for a particular mode exceeds the upper limit of the size ranges given in Table 2.1, a fraction of the particle number and mass is transferred to the adjacent mode (fractions calculated as in Eqs. 57 and 58 of Mann et al. [53]).

**Table 2.2** Characteristics of GLOMAP-mode components included in this configuration

| Component | | Summary of main sources | Density ($kg\ m^{-3}$) | Molar mass ($g\ mol^{-1}$) |
|---|---|---|---|---|
| Sulphate | SU | Volcanic eruptions, power plants, industry, road transport, shipping, domestic biofuel combustion, wildfires, marine biosphere | 1769 | 98.0 |
| Black Carbon | BC | Wildfires, fossil fuel combustion, biofuel combustion | 1500 | 12.0 |
| Particulate Organic Matter | POM | Wildfires, fossil fuel combustion, biofuel combustion, vegetation | 1500 | 16.8 |
| Sea-salt | SS | Oceans | 1600 | 58.4 |

**Table 2.3** Sulphur based reactions included in GLOMAP-mode [79]

| Reaction | References |
|---|---|
| $DMS + OH \rightarrow SO_2$ | Atkinson et al. [4] |
| $DMS + OH \rightarrow 0.6\ SO_2 + 0.4\ DMSO$ | Pham et al. [60] |
| $DMSO + OH \rightarrow 0.6\ SO_2 + 0.4\ MSA$ | Pham et al. [60] |
| $DMS + NO_3 \rightarrow SO_2$ | Atkinson et al. [4] |
| $CS_2 + OH \rightarrow SO_2 + COS$ | Pham et al. [60] |
| $COS + OH \rightarrow SO_2$ | Pham et al. [60] |
| $SO_2 + OH \rightarrow H_2SO_4$ | Pham et al. [60] |

## *2.1.2 Gas-Phase Emissions and Processes*

In this work, six-hourly mean concentrations of the hydroxyl radical (OH), ozone ($O_3$), the nitrate radical ($NO_3$), the hydroperoxy radical ($HO_2$) and hydrogen peroxide ($H_2O_2$) are prescribed from a previous TOMCAT simulation [2]. The treatment of $H_2O_2$ is semi-prognostic; in low-level clouds, it is depleted by oxidation of S(IV) (generating S(VI)), and replenished by self-reaction of $HO_2$ [33, 68].

### 2.1.2.1 Sulphur Emissions and Gas-Phase Chemistry

The sulphur chemistry included in GLOMAP-mode is detailed in Table 2.3. Phytoplankton emissions of dimethyl-sulphide (DMS) are calculated online using monthly sea-water DMS concentrations from Kettle and Andreae [35], wind-speed and sea-air gas exchange. Gas-phase sulphur dioxide ($SO_2$) emissions for the year 2000 are included from anthropogenic sources [15] and wildfires [85]; additionally, $SO_2$ from both continuous [1] and explosive [30] volcanic eruptions is included.

There are no emission inventories for carbon disulphide ($CS_2$) or carbonyl sulphide (COS). In GLOMAP-mode, anthropogenic sources of $CS_2$ and COS, such as biomass burning, automobile exhausts and chemical industries [36], are represented by a constant molar emission in proportion (0.3 and 0.08 % respectively) to anthropogenic $SO_2$ emission levels [60]. Biogenic sources of $CS_2$ and COS are represented by a constant molar emission at 1 % of the calculated DMS emission level [6].

#### 2.1.2.2 BVOC Emissions and Gas-Phase Chemistry

In the standard version of GLOMAP-mode, used in Chaps. 3–5, monthly mean emissions of monoterpenes and isoprene are taken from the Global Emissions InitiAtive (GEIA; www.geiacenter.org/) database. This inventory was compiled using the emission factors presented by Guenther et al. [26] and the vegetation distribution of Olson [58], giving total emissions of 127 Tg(C) $a^{-1}$ for monoterpenes and 503 Tg(C) $a^{-1}$ for isoprene. Figure 2.3 shows the spatial distribution of monoterpene and isoprene emissions in the GEIA database, for January and July. In the tropical regions of South America, Africa and South-east Asia, emissions of both monoterpenes and isoprene are high throughout the year. In the GEIA inventory, 78 and 87 % of the global total emission, of monoterpenes and isoprene respectively, originate from between 30°S and 30°N. In northern temperate and boreal regions, wintertime monoterpene emissions are low, and isoprene emissions are negligible (Fig. 2.3, left). In the northern hemisphere summertime (Fig. 2.3, right) emissions of BVOCs rise to a level comparable with the tropics (Fig. 2.3, right). In Chap. 6, emissions are calculated offline using the Model of Emissions of Gases and Aerosols from Nature version 2.1 (MEGAN2.1; [27]).

The amount of secondary organic material generated by the oxidation of BVOCs in the atmosphere is uncertain (Sect. 1.2.4.5). As with previous GLOMAP studies (e.g. [74, 76, 78]), and many other global-scale modelling approaches (e.g. [9, 16, 51, 61]), a fixed molar yield is applied to SOA generation from BVOC oxidation. The oxidation reaction rates used are detailed in Table 2.4; all monoterpenes are prescribed the reaction characteristics of $\alpha$-pinene, the most highly emitted compound.

As discussed in Chap. 1, the atmospheric concentration of BVOCs in any location will depend upon the relative rates of production (emission) and loss (via oxidation, mixing of air masses and deposition). Figure 2.4 shows that GLOMAP represents well the seasonal cycle in monoterpene concentration observed by Hakola et al. [29] and Lappalainen et al. [47], at a boreal forest location (Hyytiälä, Finland; 24°17′E, 61°51′N).

The standard version of GLOMAP-mode [53] includes only SOA from monoterpenes, in this work, a source of SOA from isoprene oxidation is added, following Sparcklen et al. [78]. Monoterpenes generate secondary organic material at a 13 % molar yield (after [76, 84]), and isoprene at 3 % (after [41, 42]). These yields are very uncertain and are varied in Chap. 3 as part of a sensitivity study.

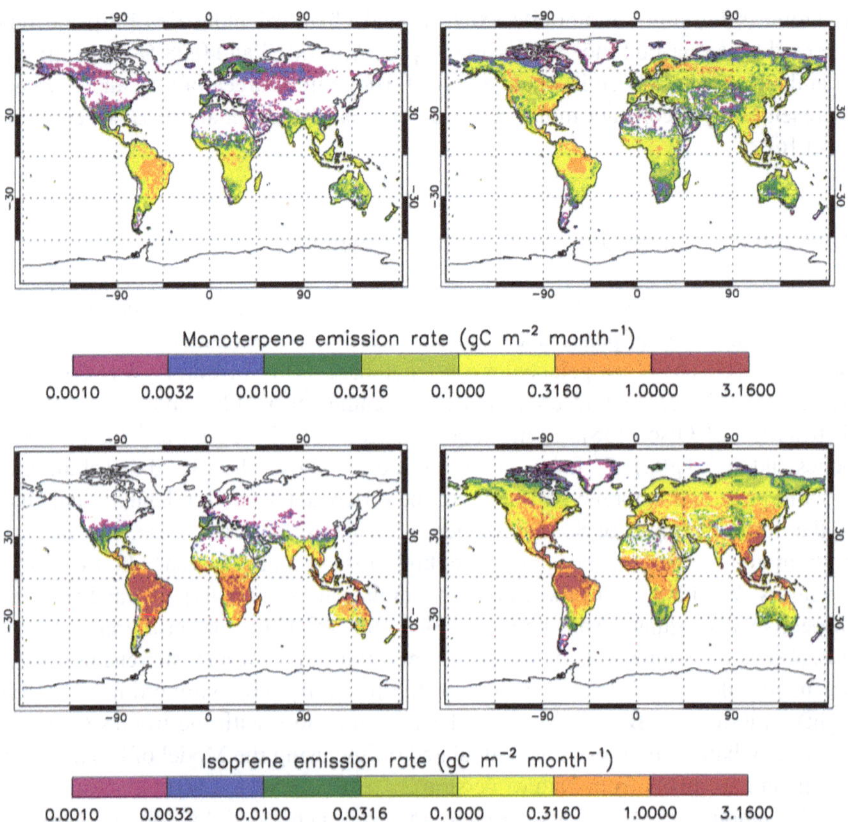


**Fig. 2.3** Monthly mean monoterpene (*upper*) and isoprene (*lower*) emission rates during January (*left*) and July (*right*) from the GEIA inventory [26]

**Table 2.4** BVOC reaction rates used in GLOMAP, taken from Atkinson et al. [4] and Atkinson et al. [3]

| Reaction | Rate coefficient ($cm^3\ s^{-1}$) |
|---|---|
| α-pinene + OH | $1.2 \times 10^{-11}$ exp (444/T) |
| α-pinene + $O_3$ | $1.01 \times 10^{-15}$ exp (−732/T) |
| α-pinene + $NO_3$ | $1.19 \times 10^{-12}$ exp (490/T) |
| Isoprene + OH | $2.7 \times 10^{-11}$ exp (390/T) |
| Isoprene + $O_3$ | $1.0 \times 10^{-14}$ exp (−1995/T) |
| Isoprene + $NO_3$ | $3.15 \times 10^{-12}$ exp (−450/T) |

### *2.1.3 Primary Particulate Emissions*

Annual mean emissions of black carbon (BC) and particulate organic matter (POM) from fossil and biofuel combustion are taken from the analysis of Bond et al. [8] and monthly wildfire emissions are from the Global Fire Emissions Database (GFEDv1)

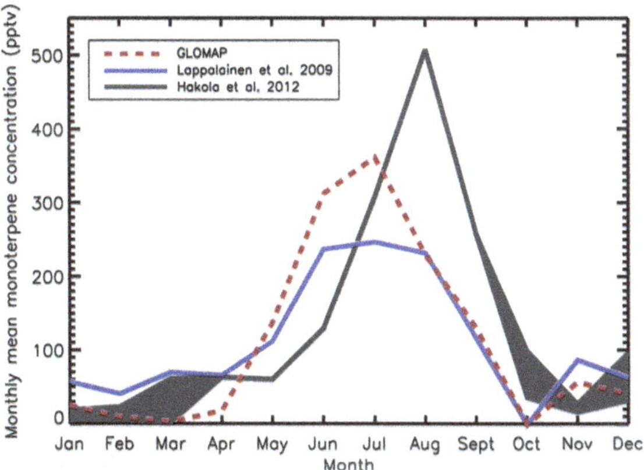


**Fig. 2.4** Seasonal cycle in monthly mean monoterpene concentration as simulated by GLOMAP-mode for the year 2000 (*dashed red line*) and observed in 2011 by Hakola et al. [29] (*grey*) and in 2006–2007 by Lappalainen et al. (47) (*blue*) at Hyytiälä, Finland

inventory [85], all represent the year 2000. Primary carbonaceous particles are emitted with the distribution characteristics described by Stier et al. [81], i.e., number median diameter $D_{ff} = 60$ nm and $D_{bf} = 150$ nm; standard deviation $\sigma_{ff} = 1.59$ and $\sigma_{bf} = 1.59$ (where *ff* = fossil fuel and *bf* = biofuel/wildfire).

Following previous work by Stier et al. [81] and the AeroCom recommendation [16], 2.5 % of all gas-phase $SO_2$ emissions are assumed to be emitted directly as particle phase sulphate to represent sub-grid scale particle formation. 50 % of all sub-grid sulphate is emitted into the accumulation mode with a number mean diameter of 150 nm and $\sigma$ of 1.59. For shipping, power plants and industrial sources, the remaining 50 % is emitted with a number mean diameter of 1.5 μm and $\sigma$ of 2.00 (i.e., into the coarse mode). For transport, domestic, wildfire and volcanic sources, the remaining 50 % is emitted with a number mean diameter of 60 nm and $\sigma$ of 1.59 (i.e., into the Aitken soluble mode).

The emission of primary sea-salt aerosol is parameterised using the sea-spray source function of [25]. Bin-resolved emissions are generated (as in [79]) and added to the accumulation (if $D_g < 1$ μm) and coarse (if $D_g > 1$ μm) modes.

## 2.1.4 Microphysical Processes

### 2.1.4.1 New Particle Formation

New particle formation, or nucleation, is known to occur throughout the atmosphere [46], from the free troposphere (e.g. [14]) to the boundary layer (e.g. [13]). Despite its ubiquity, understanding of the mechanisms driving formation of new particles in the atmosphere remains incomplete.

Several mechanisms have been proposed in order to explain atmospheric new particle formation, including binary homogeneous nucleation (BHN) of $H_2SO_4$ and $H_2O$ [43, 86], ternary nucleation of $H_2SO_4$-$H_2O$-$NH_3$ (e.g. [5, 54, 57]), and ion-induced nucleation (e.g. [48, 50, 56, 91]).

BHN is included in GLOMAP-mode and parameterised according to the hydrate-corrected, classical nucleation theory of Kulmala et al. [43]; the nucleation rate, $J_{BHN}$, is calculated as in Eq. 2.3:

$$J_{BHN} = \exp\left\{A\log\left(\frac{[H_2SO_4]}{[H_2SO_4]_{crit}}\right) + Bx_{al} + C\right\} \tag{2.3}$$

where $[H_2SO_4]_{crit}$ is the gas phase concentration of $H_2SO_4$ above which nucleation will occur and $x_{al}$ is the $H_2SO_4$ mol fraction in the critical nucleus. The coefficients $A$, $B$ and $C$ are temperature dependent and expressions for these are given in Kulmala et al. [43]. These particles are assumed to be composed of 100 sulphuric acid molecules and are added to the nucleation mode.

BHN appears able to explain nucleation rates in the free troposphere, however, new particle formation has been observed to occur in the boundary layer at far higher rates than would be predicted by BHN [79, 89], ternary [20], or ion-induced nucleation [38, 50]. Consequently, other mechanisms have been sought to explain new particle formation in the boundary layer.

Observed particle formation rates in the boundary layer appear to be proportional to the concentration of $H_2SO_4$ to the power of either 1 or 2 [44, 65, 71, 89]. In GLOMAP-mode, an empirically derived mechanism is used to represent the activation of $H_2SO_4$ clusters, as proposed by Kulmala et al. [44]. With this approach, molecular clusters (1 nm in diameter) activate at a rate, $J^*_{Act}$, calculated as in Eq. 2.4:

$$J^*_{ACT} = A[H_2SO_4] \tag{2.4}$$

The coefficient $A$ represents the complexity of the activation process and may be a function of several parameters such as temperature and the concentration of other species; in this work an empirically determined value for $A$ of $2 \times 10^{-6}\ s^{-1}$ [71] is used. In Chap. 3, several other parameterisations for the formation of the initial nucleating cluster are examined.

Following cluster activation, the production rate of particles of a measurable size, $J_m$ is calculated using the approximation of Kerminen and Kulmala [34], as given in Eq. 2.5; where $d_m$ represents the diameter of the measurable particle (taken here as 3 nm) and $d^*$ represents the cluster diameter (in this case 1 nm but values for $d^*$ vary):

$$J_m = J^* exp\left\{0.23\left[\frac{1}{d_m} - \frac{1}{d^*}\right]\frac{CS'}{GR}\right\} \tag{2.5}$$

This production rate accounts for scavenging by larger particles, and allows growth of nucleated clusters up to $d_m$ at a constant rate, *GR*, proportional to the gas phase concentration of $H_2SO_4$. The reduced condensation sink, $CS'$, is calculated as in Eq. 2.6 by integrating over the aerosol size modes, *i*, following Kulmala et al. [45]:

$$CS' = \sum_i \beta_i r_i N_i \tag{2.6}$$

Here, $\beta_i$ is the translational correction factor for the condensational mass flux [23], $r_i$ is the particle radius and $N_i$ is particle number concentration.

#### 2.1.4.2 Coagulation

Both intra- and inter-modal coagulation (i.e., particle collision) are represented in GLOMAP-mode. Particles in the soluble modes can coagulate with particles in the larger soluble and insoluble modes, whereas insoluble mode particles can coagulate only with larger insoluble mode particles. A coagulation kernel, calculated as in Spracklen [73] (equations for which are given in Sect. 2.2.6 of Mann et al. [53]), is used to determine the rate of change of particle number in each mode.

#### 2.1.4.3 Condensation and Ageing

GLOMAP-mode includes the condensation of gas-phase $H_2SO_4$ and (assumed low-volatility) secondary organic material onto existing particles. Condensing $H_2SO_4$ is added to the SU component, and condensing secondary organics are added the POM component. The rate of change of gas-phase molecular concentration of condensable material ($S_{gas}$) is calculated as in Eq. 2.7 where $C_i$ is the condensation coefficient for each mode *i* (Eq. 2.8).

$$\frac{dS_{gas}}{dt} = -\left(\sum_i C_i N_i\right) S_{gas} \tag{2.7}$$

$$C_i = 4\pi D_s \overline{r_{i,cond}} F(Kn_i) A(Kn_i) \tag{2.8}$$

$C_i$ is calculated following Fuchs and Sutugin [23] from the diffusion coefficient for $H_2SO_4$, or a typical gas-phase α-pinene oxidation product in air ($D_s$, calculated according to Fuller et al. [24] and the condensation sink radius ($\overline{r_{i,cond}}$, which is calculated as in Eq. A.1 of Mann et al. [52], based on Lehtinen et al. [49]. $C_i$ is corrected for molecular effects and limitations in interfacial mass transport by the terms $F(Kn_i)$ and $A(Kn_i)$ respectively, calculated as in Eqs. 2.9 and 2.10:

$$F(Kn_i) = \frac{1 + Kn_i}{1 + 1.71Kn_i + 1.33(Kn_i)^2} \tag{2.9}$$

$$A(Kn_i) = \frac{1 + Kn_i}{1 + 1.33Kn_i F(Kn_i)\left(\frac{1}{s} - 1\right)} \tag{2.10}$$

where $s$ is the accommodation coefficient and $Kn_i$ is the Knudsen number for each mode $i$, calculated in Eq. 2.11 using the mean free path of the relevant condensable gas in air ($MFP_{gas}$).

$$Kn_i = \frac{MFP_{gas}}{\overline{r_{i,cond}}} \tag{2.11}$$

The approach described here means that gases condense according to particle surface area; the importance of the manner in which secondary organic material is added to the existing aerosol distribution is examined in Chap. 5.

Following the condensation of soluble gas-phase species, or coagulation with smaller soluble particles, previously insoluble particles may become water soluble. Once sufficient soluble material has been accumulated, in this case ten monolayers (as in Mann et al. [53]), particles are transferred to the corresponding hydrophilic mode. This process is also known as *physical ageing*; sensitivity to the amount of soluble material required for physical ageing is examined in Chap. 3.

#### 2.1.4.4 Hygroscopic Growth

Water uptake by particles in each mode is calculated according to the *Zadanovskii-Stokes-Robinson* method (*ZSR*; Zadanovskii [92], Stokes and Robinson [82]), using data from Table B.10 in Jacobson [32] and assuming spherical particles. Organic material present in the non-hydrophilic Aitken mode is assumed to be non-hygroscopic; organic material present in any of the hydrophilic modes is either secondary or aged primary organic material and is therefore assigned a moderate hygroscopicity (65 % of the assumed water uptake for sulphate). The geometric mean wet diameter for each mode, $D_{wet_i}$, is calculated according to Eq. 2.12:

$$D_{wet_i} = \left(\frac{6}{\pi \exp(4.5\log^2\sigma_{g,i})}\sum_j V_{wet_j}\right)^{\frac{1}{3}} \tag{2.12}$$

where $V_{wet_j}$ represents the partial volume for each component and its associated water, calculated as in Eq. 2.13, where $\rho_{X,j}$ is the density of the solution of component $j$ with water (for soluble components, for insoluble component the original density is used).

$$V_{wet_j} = \frac{m_{ij}M_j}{N_a \rho_{X,j}} \tag{2.13}$$

#### 2.1.4.5 Aerosol Activation and Cloud Processing

GLOMAP-mode simulates the activation of soluble particles (with dry radius greater than $r_{act}$; here taken as 25 nm) to cloud droplets, and their subsequent growth [52, 53]. Particles in the Aitken soluble mode with dry radius greater than $r_{act}$ are transferred to the accumulation mode and sulphate produced by in cloud aqueous-phase-oxidation of $SO_2$ is then partitioned between the soluble accumulation and coarse modes.

#### 2.1.4.6 Dry Deposition

The removal of particles and gases from the atmosphere, in the absence of precipitation, is known as dry deposition. In GLOMAP-mode, dry deposition is represented using the parameterisation of Zhang et al. [93], which follows the approach of Slinn [72] in calculating a dry deposition velocity $Vel_d$, as in Eq. 2.14:

$$Vel_d = Vel_{grav} + \frac{1}{R_a + R_s} \tag{2.14}$$

where $Vel_{grav}$ is the gravitational velocity; surface resistance $R_s$ and aerodynamic resistance $R_a$ are calculated by Eqs. 2.15 and 2.16 respectively, where $k$ is the von Karman constant (equal to Eq. 2.4), $u_*$ is the surface friction velocity, $z$ is the height at which the dry deposition is being evaluated (the vertical distance from the surface), and $z_0$ is the surface roughness length:

$$R_a = \frac{1}{ku_*} \ln\left(\frac{z}{z_0}\right) \tag{2.15}$$

$$R_s = \frac{1}{3u_*(E_b + E_{im} + E_{in})} \tag{2.16}$$

The collection efficiencies associated with Brownian diffusion, impaction and interception are represented by $E_b$, $E_{im}$ and $E_{in}$ respectively, expressions for which are given in Mann et al. [53]. Experiments described in Chap. 6 involve a change to the model land-surface type, which influences dry deposition through the terms $z_0$, $u_*$, $E_b$, $E_{im}$ and $E_{in}$.

#### 2.1.4.7 Wet Deposition: Nucleation and Impaction Scavenging

Nucleation scavenging occurs when precipitation is formed in a given model level. Large-scale (dynamic) precipitation removes particles at a rate of 99.99 % conversion of cloud water to rain. The conversion rate for small-scale (convective) precipitation is calculated according to the parameterisation of Tiedtke [83], with removal of aerosol assuming a raining fraction of 30 %. In GLOMAP-mode, nucleation scavenging removes only soluble particles with a dry radius greater than $r_{scav}$ (here taken as 103 nm).

Impaction scavenging represents the removal of aerosol by falling raindrops; its implementation in GLOMAP is discussed in detail by Pringle [63]. The Marshall-Palmer raindrop size distribution as modified by Sekhon and Srivastava [70], and $D_g$ for each mode, are used to determine raindrop-particle collection efficiencies from a look-up table. An empirical relationship is used to calculate the raindrop terminal velocity [19].

### *2.1.5 Meteorological Conditions*

Meteorological fields (wind, temperature and humidity) are obtained from European Centre for Medium-Range Weather Forecasts (ECMWF) analyses by interpolation of six-hourly reanalysis (ERA-40) fields. Cloud fraction and cloud top pressure fields are taken from the International Satellite Cloud Climatology Project (ISCCP) archive (http://isccp.giss.nasa.gov/; [66]) for the year 2000. Tracer transport is controlled using the Prather [62] advection scheme, the convection scheme of Tiedtke [83] and the scheme of Holtslag and Boville [31] for boundary layer turbulence.

## 2.2 Calculation of Cloud Condensation Nuclei Concentrations

Only a subset of particles with sufficient hygroscopicity and size are able to act as cloud condensation nuclei (CCN). Following a simulation with GLOMAP-mode, CCN concentrations may be calculated offline using monthly-mean aerosol tracers, following the "κ-Köhler" approach of Petters and Kreidenweis [59], an extension to Köhler theory.

Köhler theory [39] describes the competition between the Kelvin effect (i.e., the effect of particle curvature on the vapour pressure over the surface) and the solute effect, or Raoult's Law (i.e., the contribution of each chemical solute to the vapour pressure over the particle), to predict the CCN activity of a particle with given size and composition (e.g. [69]). Building on Köhler theory, Petters and Kreidenweis

[59] introduce the term $\kappa$, a quantitative measure of water uptake ability and CCN activity, also known as the *hygroscopicity parameter*.

In this work, a $\kappa$ value is assigned to each component: SU (0.61, assuming ammonium sulphate), SS (1.28), BC (0.0) and POM (0.1), following Petters and Kreidenweis [59]. A multicomponent, $\kappa_{multi}$, is then obtained by weighting the individual $\kappa$ values by the volume fraction of each component. Whilst there is considerable uncertainty associated with the hygroscopicity parameter for organic material in the atmosphere due to the wide range of solubilities observed, $\kappa$ values close to 0.1 have been reported for secondary organic components [18, 21, 22, 28, 37], and the entire organic fraction [11, 88].

Here, CCN concentrations are calculated at a fixed uniform supersaturation of 0.2 %. This would be equivalent to an activation dry diameter of approximately 80 nm (assuming a composition of pure ammonium sulphate).

## 2.3 Suitability of GLOMAP-Mode for This Work

As a two-moment scheme, GLOMAP is able to calculate aerosol number as well as mass. This confers considerable advantage over mass-only models when examining processes that affect the evolution of the aerosol size distribution. As demonstrated by Bellouin et al. [7], the addition of particle-phase material in a mass-only model results in an increase in aerosol number. However, if both mass and number are tracked, the addition of particle-phase material (for example the condensation of secondary organics) may grow existing particles; altering their size but not necessarily their number. Particle number concentrations will however be modified by changes to the condensation and coagulation sinks resulting from the particle growth; for example, enhanced condensation of nucleating vapours due to an increase in the surface area available for condensation. Accounting for particle mass and number is particularly important if one wishes to determine the number of particles of a certain size, for example when calculating CCN concentrations.

## 2.4 Comparison to Observations

Particle and speciated mass concentrations simulated by GLOMAP-mode have been compared to observations in several previous studies (e.g. [52, 53, 67, 78, 90]) and in Chap. 3 of this thesis.

Mann et al. [53] found that GLOMAP-mode captures the spatial variability in observed POM mass concentrations over North America (Pearson's correlation coefficient, $R$, of 0.82); but that the POM mass burden was underestimated by the model (normalised mean bias, $NMB$, of $-0.57$). Spracklen et al. [78] found that GLOMAP-mode underestimated organic mass when compared against aerosol mass spectrometer observations at over 30 locations; the bias against observations was minimised when a large source of SOA (i.e., 100 Tg(SOA) $a^{-1}$) was included in the model.

CCN concentrations calculated offline (Sect. 2.2) from both GLOMAP-mode and GLOMAP-bin have been compared to the global dataset of CCN observations compiled by Spracklen et al. [77]. Spracklen et al. [77] and Schmidt et al. [67] found that both versions of GLOMAP tend to underestimate CCN concentrations when compared to the global dataset (e.g. NMB = −38 %; Schmidt et al. [67], NMB = −25 %; Spracklen et al. [77]). Spracklen et al. [77] demonstrated that simulated CCN concentrations were particularly sensitive to the treatment and atmospheric processing of primary particulate emissions, and the new particle formation mechanisms used; this will be explored in Chap. 3.

## References


1. Andres RJ, Kasgnoc AD (1998) A time-averaged inventory of subaerial volcanic sulfur emissions. J Geophys Res 103(D19):25251–25261
2. Arnold SR et al (2005) A three-dimensional model study of the effect of new temperature-dependent quantum yields for acetone photolysis. J Geophys Res 110(D22):D22305
3. Atkinson R et al (2006) Evaluated kinetic and photochemical data for atmospheric chemistry: volume II—gas phase reactions of organic species. Atmos Chem Phys 6(11):3625–4055
4. Atkinson R et al (1989) Evaluated kinetic and photochemical data for atmospheric chemistry: supplement III. IUPAC subcommittee on gas kinetic data evaluation for atmospheric chemistry. J Phys Chem Ref Data 18:881–1097
5. Ball SM et al (1999) Laboratory studies of particle nucleation: Initial results for $H_2SO_4$, $H_2O$, and $NH_3$ vapors. J Geophys Res Atmos 104(D19):23709–23718
6. Bates TS et al (1992) Sulfur emissions to the atmosphere from natural sourees. J Atmos Chem 14(1–4):315–337
7. Bellouin N et al (2013) Impact of the modal aerosol scheme GLOMAP-mode on aerosol forcing in the Hadley Centre Global Environmental Model. Atmos Chem Phys 13(6):3027–3044
8. Bond TC et al (2004) A technology-based global inventory of black and organic carbon emissions from combustion. J Geophys Res 109(D14):D14203
9. Boy M et al (2006) MALTE—model to predict new aerosol formation in the lower troposphere. Atmos Chem Phys 6(12):4499–4517
10. Browse J et al (2012) The scavenging processes controlling the seasonal cycle in arctic sulphate and black carbon aerosol. Atmos Chem Phys 12(15):6775–6798
11. Chang RYW et al (2010) The hygroscopicity parameter (κ) of ambient organic aerosol at a field site subject to biogenic and anthropogenic influences: relationship to degree of aerosol oxidation. Atmos Chem Phys 10(11):5047–5064
12. Chipperfield MP (2006) New version of the TOMCAT/SLIMCAT off-line chemical transport model: Intercomparison of stratospheric tracer experiments. Q J R Meteorol Soc 132(617):1179–1203
13. Clarke AD et al (1998) Particle nucleation in the tropical boundary layer and its coupling to marine sulfur sources. Science 282(5386):89–92
14. Clarke AD et al (1999) Nucleation in the equatorial free troposphere: favorable environments during PEM-Tropics. J Geophys Res Atmos 104(D5):5735–5744

15. Cofala J et al (2005) Scenarios of world anthropogenic emissions of $SO_2$, $NO_x$ and CO up to 2030. In Internal report of the transboundary air pollution programme, International Institute for Applied Systems Analysis. Laxenburg, Austria
16. Dentener F et al (2006) Emissions of primary aerosol and precursor gases in the years 2000 and 1750 prescribed data-sets for AeroCom. Atmos Chem Phys 6(12):4321–4344
17. Dunne EM et al (2012) No statistically significant effect of a short-term decrease in the nucleation rate on atmospheric aerosols. Atmos Chem Phys 12(23):11573–11587
18. Dusek U et al (2010) Enhanced organic mass fraction and decreased hygroscopicity of cloud condensation nuclei (CCN) during new particle formation events. Geophys Res Lett 37(3):L03804
19. Easter RC, Hales JM (1983) Interpretation of the OSCAR data for reactive gas scavenging. In: Precipitation scavenging, dry deposition and resuspension. Elsevier, New York, pp 649–662
20. Elleman RA, Covert DS (2009) Aerosol size distribution modeling with the community multiscale air quality modeling system in the Pacific Northwest: 2. Parameterizations for ternary nucleation and nucleation mode processes. J Geophys Res Atmos 114(D11):D11207
21. Engelhart GJ et al (2008) CCN activity and droplet growth kinetics of fresh and aged monoterpene secondary organic aerosol. Atmos Chem Phys 8(14):3937–3949
22. Engelhart GJ et al (2011) Cloud condensation nuclei activity of isoprene secondary organic aerosol. J Geophys Res 116(D2):D02207
23. Fuchs NA, Sutugin AG (1971) Highly dispersed aerosols. In: Topics in current aerosol research. Pergamin, New York, pp 1–60
24. Fuller EN et al (1966) New method for prediction of binary gas-phase diffusion coefficients. Ind Eng Chem 58(5):18–27
25. Gong SL (2003) A parameterization of sea-salt aerosol source function for sub- and super-micron particles. Global Biogeochem Cycles 17(4):1097
26. Guenther A et al (1995) A global model of natural volatile organic compound emissions. J Geophys Res 100(D5):8873–8892
27. Guenther AB et al (2012) The model of emissions of gases and aerosols from nature version 2.1 (MEGAN2.1): an extended and updated framework for modeling biogenic emissions. Geosci Model Dev 5(6):1471–1492
28. Gunthe SS et al (2009) Cloud condensation nuclei in pristine tropical rainforest air of Amazonia: size-resolved measurements and modeling of atmospheric aerosol composition and CCN activity. Atmos Chem Phys 9(19):7551–7575
29. Hakola H et al (2012) In situ measurements of volatile organic compounds in a boreal forest. Atmos Chem Phys 12(23):11665–11678
30. Halmer MM et al (2002) The annual volcanic gas input into the atmosphere, in particular into the stratosphere: a global data set for the past 100 years. J Volcanol Geoth Res 115(3–4):511–528
31. Holtslag AAM, Boville BA (1993) Local versus nonlocal boundary-layer diffusion in a global climate model. J Clim 6(10):1825–1842
32. Jacobson M (2005) Fundamentals of atmospheric modelling. Cambridge University Press, New York
33. Jones A et al (2001) Indirect sulphate aerosol forcing in a climate model with an interactive sulphur cycle. J Geophys Res Atmos 106(D17):20293–20310
34. Kerminen V-M, Kulmala M (2002) Analytical formulae connecting the "real" and the "apparent" nucleation rate and the nuclei number concentration for atmospheric nucleation events. J Aerosol Sci 33(4):609–622
35. Kettle AJ, Andreae MO (2000) Flux of dimethylsulfide from the oceans: a comparison of updated data sets and flux models. J Geophys Res Atmos 105(D22):26793–26808
36. Khalil MAK, Rasmussen RA (1984) Global sources, lifetimes and mass balances of carbonyl sulfide (OCS) and carbon disulfide (CS2) in the earth's atmosphere. Atmos Environ (1967), 18(9):1805–1813

37. King SM et al (2010) Cloud droplet activation of mixed organic-sulfate particles produced by the photooxidation of isoprene. Atmos Chem Phys 10(8):3953–3964
38. Kirkby J et al (2011) Role of sulphuric acid, ammonia and galactic cosmic rays in atmospheric aerosol nucleation. Nature 476(7361):429–433
39. Köhler H (1936) The nucleus in and the growth of hygroscopic droplets. Trans Faraday Soc 32:1152–1161
40. Korhonen H et al (2010) Enhancement of marine cloud albedo via controlled sea spray injections: a global model study of the influence of emission rates, microphysics and transport. Atmos Chem Phys 10(9):4133–4143
41. Kroll JH et al (2005) Secondary organic aerosol formation from isoprene photooxidation under high-NOx conditions. Geophys Res Lett 32(18):L18808
42. Kroll JH et al (2006) Secondary organic aerosol formation from isoprene photooxidation. Environ Sci Technol 40(6):1869–1877
43. Kulmala M et al (1998) Parameterisations for sulphuric acid/water nucleation rates. J Geophys Res Atmos 103(D7):8301–8307
44. Kulmala M et al (2006) Cluster activation theory as an explanation of the linear dependence between formation rate of 3 nm particles and sulphuric acid concentration. Atmos Chem Phys 6:787–793
45. Kulmala M et al (2001) On the formation, growth and composition of nucleation mode particles. Tellus B 53(4):479–490
46. Kulmala M et al (2004) Formation and growth rates of ultrafine atmospheric particles: a review of observations. J Aerosol Sci 35(2):143–176
47. Lappalainen HK et al (2009) Day-time concentrations of biogenic volatile organic compounds in a boreal forest canopy and their relation to environmental and biological factors. Atmos Chem Phys 9(15):5447–5459
48. Lee S-H et al (2003) Particle formation by ion nucleation in the upper troposphere and lower stratosphere. Science 301(5641):1886–1889
49. Lehtinen KEJ et al (2003) On the concept of condensation sink diameter. Boreal Environ Res 8(4):405–411
50. Lovejoy ER et al (2004) Atmospheric ion-induced nucleation of sulfuric acid and water. J Geophys Res Atmos 109(D8):D08204
51. Makkonen R et al (2009) Sensitivity of aerosol concentrations and cloud properties to nucleation and secondary organic distribution in ECHAM5-HAM global circulation model. Atmos Chem Phys 9(5):1747–1766
52. Mann GW et al (2012) Intercomparison of modal and sectional aerosol microphysics representations within the same 3-D global chemical transport model. Atmos Chem Phys 12(10):4449–4476
53. Mann GW et al (2010) Description and evaluation of GLOMAP-mode: a modal global aerosol microphysics model for the UKCA composition-climate model. Geosci. Model Dev. 3(2):519–551
54. Merikanto J et al (2007) New parameterization of sulfuric acid-ammonia-water ternary nucleation rates at tropospheric conditions. J Geophys Res Atmos 112(D15):D15207
55. Merikanto J et al (2009) Impact of nucleation on global CCN. Atmos Chem Phys 9:8601–8616
56. Modgil MS et al (2005) A parameterization of ion-induced nucleation of sulphuric acid and water for atmospheric conditions. J Geophys Res Atmos 110(D19):D19205
57. Napari I et al (2002) An improved model for ternary nucleation of sulfuric acid–ammonia–water. J Chem Phys 116(10):4221–4227
58. Olson J (1992) World ecosystems (W E1.4): Digital raster data on a 10 min geographic 1,080 × 2,160 grid. NOAA National Geophysical Data Center, Boulder
59. Petters MD, Kreidenweis SM (2007) A single parameter representation of hygroscopic growth and cloud condensation nucleus activity. Atmos Chem Phys 7(8):1961–1971
60. Pham M et al (1995) A three-dimensional study of the tropospheric sulfur cycle. J Geophys Res Atmos 100(D12):26061–26092

61. Pierce JR et al (2013) Weak global sensitivity of cloud condensation nuclei and the aerosol indirect effect to Criegee + $SO_2$ chemistry. Atmos Chem Phys 13(6):3163–3176
62. Prather MJ (1986) Numerical advection by conservation of second-order moments. J Geophys Res Atmos 91(D6):6671–6681
63. Pringle KJ (2006) Aerosol-cloud interactions in a global model of aerosol microphysics, Ph.D. thesis, University of Leeds
64. Reddington CL et al (2011) Primary versus secondary contributions to particle number concentrations in the European boundary layer. Atmos Chem Phys 11(23):12007–12036
65. Riipinen I et al (2007) Connections between atmospheric sulphuric acid and new particle formation during QUEST III-IV campaigns in Heidelberg and Hyytiälä. Atmos Chem Phys 7(8):1899–1914
66. Rossow WB, Schiffer RA (1999) Advances in Understanding Clouds from ISCCP. B Am Meteorol Soc 80(11):2261–2287
67. Schmidt A et al (2012) Importance of tropospheric volcanic aerosol for indirect radiative forcing of climate. Atmos Chem Phys 12(16):7321–7339
68. Seinfeld JH, Pandis SN (2006a) Chemistry of the atmospheric aqueous phase. In: Atmospheric chemistry and physics: from air pollution to climate change, 2nd edn. Wiley, pp 284–349
69. Seinfeld JH, Pandis SN (2006b). Cloud physics. In: Atmospheric chemistry and physics: from air pollution to climate change, 2nd edn. Wiley, pp 761–822
70. Sekhon RS, Srivastava RC (1971) Doppler radar observations of drop-size distributions in a thunderstorm. J Atmos Sci 28(6):983–994
71. Sihto S-L et al (2006) Atmospheric sulphuric acid and aerosol formation: implications from atmospheric measurements for nucleation and early growth mechanisms. Atmos Chem Phys 6:4079–4091
72. Slinn WGN (1982) Predictions for particle deposition to vegetative canopies. Atmos Environ (1967), 16(7):1785–1794
73. Spracklen DV (2005) Development and application of a global model of aerosol processes, Ph.D. thesis, University of Leeds, UK
74. Spracklen DV et al (2008) Boreal forests, aerosols and the impacts on clouds and climate. Philos Trans R Soc A 366:4613–4626
75. Spracklen DV et al (2008) Contribution of particle formation to global cloud condensation nuclei concentrations. Geophys Res Lett 35(6):L06808
76. Spracklen DV et al (2006) The contribution of boundary layer nucleation events to total particle concentrations on regional and global scales. Atmos Chem Phys 6(12):5631–5648
77. Spracklen DV et al (2011) Global cloud condensation nuclei influenced by carbonaceous combustion aerosol. Atmos Chem Phys 11(17):9067–9087
78. Spracklen DV et al (2011) Aerosol mass spectrometer constraint on the global secondary organic aerosol budget. Atmos Chem Phys 11(23):12109–12136
79. Spracklen DV et al (2005) A global off-line model of size-resolved aerosol microphysics: I. Model development and prediction of aerosol properties. Atmos Chem Phys 5(8):2227–2252
80. Spracklen DV et al (2005) A global off-line model of size-resolved aerosol microphysics: II. Identification of key uncertainties. Atmos Chem Phys 5(12):3233–3250
81. Stier P et al (2005) The aerosol-climate model ECHAM5-HAM. Atmos Chem Phys 5(4):1125–1156
82. Stokes RH, Robinson RA (1966) Interactions in aqueous nonelectrolyte solutions. I. Solute-solvent equilibria. J Phys Chem 70(7):2126–2131
83. Tiedtke M (1989) A comprehensive mass flux scheme for cumulus parameterization in large-scale models. Mon Weather Rev 117(8):1779–1800
84. Tunved P et al (2004) A pseudo-Lagrangian model study of the size distribution properties over Scandinavia: transport from Aspvreten to Värriö. Atmos Chem Phys Discuss 4(6):7757–7794
85. van der Werf GR et al (2004) Continental-scale partitioning of fire emissions during the 1997 to 2001 El Niño/La Niña period. Science 303(5654):73–76

86. Vehkamäki H et al (2002) An improved parameterization for sulfuric acid–water nucleation rates for tropospheric and stratospheric conditions. J Geophys Res Atmos 107(D22):4622
87. Vignati E et al (2004) M7: an efficient size-resolved aerosol microphysics module for large-scale aerosol transport models. J Geophys Res 109(D22):D22202
88. Wang J et al (2008) Effects of aerosol organics on cloud condensation nucleus (CCN) concentration and first indirect aerosol effect. Atmos Chem Phys 8(21):6325–6339
89. Weber RJ et al (1996) Measured atmospheric new particle formation rates: implications for nucleation mechanisms. Chem Eng Commun 151(1):53–64
90. Woodhouse MT et al (2010) Low sensitivity of cloud condensation nuclei to changes in the sea-air flux of dimethyl-sulphide. Atmos Chem Phys 10(16):7545–7559
91. Yu F, Turco RP (2001) From molecular clusters to nanoparticles: Role of ambient ionization in tropospheric aerosol formation. J Geophys Res Atmos 106(D5):4797–4814
92. Zadanovskii AB (1948) New methods for calculating solubilities of electrolytes in multicomponent systems. Zhurnal fizicheskoi khimii (Russ J Phys Chem) 22:1475–1485
93. Zhang L et al (2001) A size-segregated particle dry deposition scheme for an atmospheric aerosol module. Atmos Environ 35(3):549–560

# Chapter 3
# The Impact of Biogenic SOA on Particle and Cloud Condensation Nuclei Concentration

## 3.1 Introduction

As described in Chap. 1, the presence of biogenic SOA affects the number and size of particles in the atmosphere. Organic oxidation products may condense onto existing particles and aid their growth to larger sizes (e.g., [30]), enhance particle solubility [27], and contribute to new particle formation (e.g., [25]).

In this chapter, the role of biogenic secondary organic aerosol in the atmosphere is quantified using GLOMAP-mode. A set of experiments were conducted to determine the processes and parameters to which the impacts of biogenic SOA are most sensitive, and simulated particle concentrations were compared to a range of observations.

## 3.2 Experimental Design

To examine the extent to which the simulated climate impacts of biogenic SOA are sensitive to uncertainties in aerosol processes, the series of model experiments described in Table 3.1 were completed. These simulations explore the impact of uncertainty in the SOA yield from BVOC oxidation, the role of organic oxidation products in new particle formation, and the interaction between SOA and primary carbonaceous aerosol emissions. All simulations are performed for the year 2000, with 6 months spin-up from zero initial aerosol (i.e., 18 months total simulation).

### *3.2.1 Yield*

In the standard configuration (Expt. 2 to 4) monoterpenes and isoprene generate a condensable secondary organic material at fixed molar yields of 13 % (following



C.E. Scott, *The Biogeochemical Impacts of Forests and the Implications for Climate Change Mitigation*, Springer Theses, DOI 10.1007/978-3-319-07851-9_3

**Table 3.1** Summary of simulations performed; results from Expt. 21–27 are discussed in Chap. 4

| Exp. No. | Expt. name | Description | BVOCs included | Global production of SOA (Tg(SOA) $a^{-1}$) |
|---|---|---|---|---|
| 1 | ACT | BHN with Eq. 3.1 within the boundary layer | None | 0 |
| 2 | ACT_m | | Monoterpenes | 20.4 |
| 3 | ACT_i | | Isoprene | 16.6 |
| 4 | ACT_mi | | Mono + Iso | 37.0 |
| 5 | ACT_mi_x0.5 | 0.5 × SOA production yield | Mono + Iso | 18.5 |
| 6 | ACT_mi_x2 | 2 × SOA production yield | Mono + Iso | 74.0 |
| 7 | ACT_mi_x5 | 5 × SOA production yield | Mono + Iso | 185.1 |
| 8 | ACT_mi_noSOAage | No transfer of non-hydrophilic particles to the hydrophilic distribution via condensation of secondary organics | Mono + Iso | 37.0 |
| 9 | ACT_fast_age | One soluble monolayer required to transfer non-hydrophilic particles to the hydrophilic distribution | None | 0 |
| 10 | ACT_mi_fast_age | | Mono + Iso | 37.0 |
| 11 | ACT_BCOCsmall | Size distribution of primary BC and OC emissions set to AeroCom recommendation [3] | None | 0 |
| 12 | ACT_mi_BCOCsmall | | Mono + Iso | 37.0 |
| 13 | BHN | Binary homogeneous nucleation only | None | 0 |
| 14 | BHN_m | | Monoterpenes | 20.4 |
| 15 | Org1 | BHN with Eq. 3.2 throughout the atmosphere | None | 0 |
| 16 | Org1_m | | Monoterpenes | 20.4 |
| 17 | Org2 | BHN with Eq. 3.3 throughout the atmosphere | None | 0 |
| 18 | Org2_m | | Monoterpenes | 20.4 |
| 19 | Org3 | BHN with Eq. 3.4 throughout the atmosphere | None | 0 |
| 20 | Org3_m | | Monoterpenes | 20.4 |
| 21 | ACT_1750 | 1750 emissions of BC, POM and $SO_2$ taken from [3] | None | 0 |
| 22 | ACT_1750_mi | | Mono + Iso | 37.0 |
| 23 | ACT_1750_mi_x0.5 | 1750 emissions, with 0.5 × SOA production yield | Mono + Iso | 18.5 |
| 24 | ACT_1750_mi_x2 | 1750 emissions, with 2 × SOA production yield | Mono + Iso | 74.0 |
| 25 | ACT_1750_mi_x5 | 1750 emissions, with 5 × SOA production yield | Mono + Iso | 185.1 |
| 26 | Org1_1750 | 1750 emissions, standard yields and BHN with Eq. 3.2 throughout the atmosphere | None | 0 |
| 27 | Org1_1750_m | | Monoterpenes | 20.4 |

[33, 34, 41]) and 3 % [16, 17], respectively. These yields are very uncertain, so to account for this, experiments were conducted in which the SOA production yields were multiplied by a factor of 0.5, 2 and 5 (Expt. 5–7), resulting in a global production of between 18.5 Tg(SOA) $a^{-1}$ and 185.0 Tg(SOA) $a^{-1}$. The global production of biogenic SOA from each experiment is detailed in Table 3.1 and all lie within the wide range of previous estimates (Sect. 1.2.4.5; [4, 5, 9, 8, 11, 37]).

### 3.2.2 New Particle Formation

New particle formation has been shown to strongly affect CCN concentrations (e.g., [24, 35]). To explore the potential role of BVOC oxidation products in the formation of new particles, the impact of biogenic SOA is quantified using five different representations of new particle formation (Table 3.1). All experiments include the binary homogeneous nucleation (BHN) of sulphuric acid and water [18] which occurs mainly in the free troposphere (e.g., [32]). However, as discussed in Chap. 2, observed particle formation rates in the boundary layer cannot be explained by binary nucleation of $H_2SO_4$ and $H_2O$ alone.

Here, BHN is combined with four parameterisations for the formation of new particles in the boundary layer (Table 3.2); ACT, Org1 and Org2 are taken from existing literature, whilst Org3 is evaluated for the first time here.

The first additional boundary layer mechanism (ACT; Eq. 3.1) is described in Chap. 2 and is based on the activation of $H_2SO_4$ clusters, at a rate ($J^*$) proportional to the gas phase concentration of $H_2SO_4$. This mechanism is restricted to the boundary layer and BHN is allowed to proceed at higher altitudes.

Sulphuric acid and organic compounds have both been implicated in the appearance of new particles at an observable size, but the identity of compounds initiating nucleation is inherently difficult to establish due to the practical limitations in measuring the composition of the smallest particles. In smog chamber experiments, [25] found that the formation rate of 1.5 nm clusters was proportional to the product of the gas-phase concentrations of $H_2SO_4$ and low volatility products from the photo-oxidation of SOA precursor species, in this case 1,3,5-trimethylbenzene (Org1; Eq. 3.2). Paasonen et al. [26] found good correlation against a wide observational dataset when the particle formation rate combined a term based on the product of the concentrations of $H_2SO_4$ and an organic molecule, and a kinetic nucleation term proportional to the square of $H_2SO_4$ concentration (Org2; Eq. 3.3). In Eqs. 3.2 and 3.3, the term *OxOrg* represents the products of monoterpene oxidation only, since the role of isoprene oxidation products in new particle formation remains unclear [12, 13].

**Table 3.2** Summary of new particle formation mechanisms used during this chapter

| Mechanism | | Cluster formation rate ($J^*$) | Coefficient values ($s^{-1}$) | Cluster size (nm) | |
|---|---|---|---|---|---|
| ACT | [21] | $A[H_2SO_4]$ | $A = 2 \times 10^{-6}$ | 0.8 | (3.1) |
| Org1 | [25] | $k_1[H_2SO_4][OxOrg]$ | $k_1 = 5 \times 10^{-13}$ | 1.5 | (3.2) |
| Org2 | [26] | $k_2[H_2SO_4]^2 + k_3[H_2SO_4][OxOrg]$ | $k_2 = 1.1 \times 10^{-14}$<br>$k_3 = 3.2 \times 10^{-14}$ | 2.0 | (3.3) |
| Org3 | [29] | $k_4[H_2SO_4]^2[BioOxOrg]$ | $k_4 = 5.5 \times 10^{-21}$ | 1.7 | (3.4) |

#### 3.2.2.1 The CLOUD Mechanism for New Particle Formation

The fourth additional mechanism (Org3; Eq. 3.4) for new particle formation is the result of experiments performed in the CLOUD (Cosmics Leaving Outdoor Droplets) chamber at CERN. The CLOUD chamber is a stainless steel vessel of 26 $m^3$, in which nucleation experiments may be conducted under stable and contaminant free conditions [14]. A new parameterisation was developed to describe the dependence of nucleation rates on the concentration of $H_2SO_4$ and organic oxidation products (*BioOxOrg*) in the chamber. *BioOxOrg* represents the oxidation products of pinanediol (*PD*, $C_{10}H_{18}O_2$), chosen as a model compound for the first generation oxidation products of monoterpenes. The effects of $H_2SO_4$ and *BioOxOrg* on the nucleation rate were determined experimentally by independently varying their concentrations in the CLOUD chamber; see [29] and its Supplementary Material for updated values and further details on the experimental setup and protocol.

The formation rate of 1.7 nm clusters was found to be proportional to the concentration of $H_2SO_4$ to the power 2, and the concentration of *BioOxOrg* to the power 1, with a multi-component pre-factor ($k_4$; Table 3.2) of $5.5 \times 10^{-21}$ $s^{-1}$. The concentration of *BioOxOrg* in the chamber, under steady state conditions, may be described as in Eq. 3.5, in relation to its production (via oxidation of PD by OH) and loss rates (i.e., vapour loss to the chamber walls ($k_{wall}$), condensation onto existing particles ($k_{cond}$), and dilution due to replacement of sampled air in the chamber with clean air ($k_{dil}$)). The derivation of these loss terms is described in [29]. Here, it is sufficient to note that $k_{wall}$ is the dominant loss rate.

$$[BioOxOrg] = \frac{k_{PD,OH}[PD][OH]}{(k_{wall} + k_{cond} + k_{dil})} \tag{3.5}$$

Similarly, the rate of change of PD concentration may be described as in Eq. 3.6, in relation to its rate of production (via oxidation of α-pinene) and loss (via oxidation to *BioOxOrg*).

$$\frac{d[PD]}{dt} = k_{\alpha pin,OH}[\alpha pin][OH] - k_{PD,OH}[PD][OH] \tag{3.6}$$

Therefore, at steady-state, when $d$[PD]/$dt$ is zero, the concentration of PD may be expressed as in Eq. 3.7

$$[PD] = \frac{(k_{\alpha pin,OH}[\alpha pin])}{k_{PD,OH}} \tag{3.7}$$

For implementation in GLOMAP-mode, Eqs. 3.4, 3.5, 3.6 and 3.7 were combined to give Eq. 3.8, where $k_{CS}$, the atmospheric condensation sink (i.e., rate of condensation onto existing particles, $s^{-1}$) replaces the loss term $k_{wall}$, and is calculated assuming the diffusion characteristics of a typical α-pinene oxidation product (see Appendix A1 of [23]).

$$J^* = k_4[H_2SO_4]^2\left[\frac{(k_{\alpha pin,OH}[\alpha pin][OH])}{k_{CS}}\right] \tag{3.8}$$

### 3.2.3 Characteristics of Primary Carbonaceous Emissions

In GLOMAP, the simulated aerosol size distributions and therefore CCN concentrations, are sensitive to the treatment of primary emissions [28, 38], in particular the emission characteristics of primary carbonaceous aerosol. As described in Chap. 2, primary carbonaceous particles are emitted with the distribution characteristics described by Stier et al. [39]. Sensitivity to this choice, in terms of the impact of biogenic SOA, is explored with an additional set of simulations (*ACT_BCOCsmall*) using the smaller emission size (i.e., number median diameter $D_{ff} = 30$ nm, $D_{bf} = 80$ nm; where *ff* = fossil fuel and *bf* = biofuel/wildfire respectively), but wider distribution (i.e., standard deviation $\sigma_{ff} = 1.8$, $\sigma_{bf} = 1.8$) recommended by AeroCom [3].

The process of physical ageing, by which non-hydrophilic particles become hydrophilic via condensation of soluble material, is poorly understood [11]. In GLOMAP, primary BC/OC particles are emitted into a non-hydrophilic distribution and are transferred to an internally mixed hydrophilic distribution after the condensation of a specific amount of condensable material. Here, the standard simulations assume that ten monolayers of condensable material (secondary organics or $H_2SO_4$) are required to sufficiently coat an insoluble particle for it to become soluble. To examine the importance of this process, and its representation with respect to the impact of biogenic SOA, an additional simulation in which physical ageing does not occur as a result of the condensation of secondary organics (*ACT_mi_noSOAage*) is performed, and another in which only one soluble monolayer is required (*ACT_mi_fast_age*).

### *3.2.4 Presence of Anthropogenic Emissions*

To test the sensitivity of the impact of biogenic SOA to the presence of anthropogenic emissions, a set of simulations (Expts. 21 to 27) are conducted which include anthropogenic emissions from the year 1750 (BC and POM from wildfire and biofuel, $SO_2$ from wildfire and domestic sources), compiled for the AeroCom initiative [3]. In the 1750 emission dataset, wildfire emissions are scaled versions of monthly mean data from 1998–2002 (scaled according to population ratio (i.e., 1750 vs. 1990) from the 100 Year database for Integrated Environmental Assessments (HYDE; www.rivm.nl/hyde)). Emissions from deforestation fires are scaled by population, whereas emissions from other land surfaces (grassland, shrub/bush, agricultural) are 60 % scaled by population, as the remaining 40 % is assumed to burn regardless of human activity; forest emissions in high latitudes (Europe, North America and Russia) are doubled from current emissions to reflect the fact that there would have been less wildfire suppression in the past. Consequently, tropical deforestation emissions are lower in the 1750 dataset, but boreal wildfire emissions are higher [3]. Results from these simulations are discussed in Chap. 4.

## 3.3 Results

### *3.3.1 Changes to Total Particle Concentration*

Table 3.3 reports the impact of biogenic SOA on simulated aerosol properties. The impact of biogenic SOA on global annual mean total particle number concentration (greater than 3 nm dry diameter; $N_3$) depends upon the nucleation mechanism implemented in the model. Whilst the condensation of secondary organic material leads to particle growth, this increase in particle size also enhances the condensation sink for nucleating gases ($H_2SO_4$ and organic oxidation products) and the coagulational sink for newly formed particles. As a result, the global annual mean $N_3$ concentration is reduced by 7.9 % when monoterpene emissions are included with activation boundary layer nucleation (*ACT_m*) and by 0.4 % when BHN is the only new particle formation mechanism (*BHN_m*). In contrast, when organics contribute directly to nucleation (*Org1_m*, *Org2_m* and *Org3_m*), a global annual mean increase in $N_3$ of between 22.0 and 142.0 % is simulated. In these simulations, the additional nucleation resulting from the presence of organics outweighs the moderate reduction in $N_3$ due to the enhanced condensation sink.

**Table 3.3** Global annual mean surface-level changes to $N_3$ (number concentration of particles >3 nm diameter) and CCN (0.2 % supersaturation) concentration, relative to an equivalent control simulation including no BVOC emission

| Exp. No. | Expt. name | $\Delta N_3$ ($cm^{-3}$) | $\Delta$CCN ($cm^{-3}$) |
|---|---|---|---|
| 2 | ACT_m | −64.3 (−7.9 %) | +23.5 (+10.4 %) |
| 3 | ACT_i | −43.1 (−5.3 %) | +19.3 (+8.5 %) |
| 4 | ACT_mi | −79.6 (−9.7 %) | +29.0 (+12.8 %) |
| 5 | ACT_mi_x0.5 | −56.7 (−6.9 %) | +22.1 (+9.7 %) |
| 6 | ACT_mi_x2 | −106.7 (−13.0 %) | +36.7 (+16.2 %) |
| 7 | ACT_mi_x5 | −145.5 (−17.8 %) | +48.0 (+21.1 %) |
| 8 | ACT_mi_noSOAage | −81.3 (−9.9 %) | +8.3 (+3.6 %) |
| 10 | ACT_mi_fast_age | −81.5 (−9.9 %) | +21.8 (+8.9 %) |
| 12 | ACT_mi_BCOCsmall | −55.8 (−5.0 %) | +46.0 (+17.5 %) |
| 14 | BHN_m | −1.7 (−0.4 %) | +20.3 (+10.4 %) |
| 16 | Org1_m | +643.0 (+142 %) | +88.1 (+45.2 %) |
| 18 | Org2_m | +188.0 (+22.0 %) | +39.0 (+15.5 %) |
| 20 | Org3_m | +302.3 (+66.7 %) | +69.9 (+35.8 %) |

Each global annual mean change is expressed as a percentage, relative to the control, in brackets

### 3.3.2 Changes to Cloud Condensation Nuclei Concentrations

In all the model configurations examined here, biogenic SOA increases the simulated global annual mean surface CCN concentration, with the relative enhancement ranging between 3.6 and 45.2 % (Table 3.1). The spatial distribution of changes to CCN concentration (Fig. 3.1) does not simply match the distribution of BVOC emissions (Fig. 2.3) due to the diverse range of processes controlling the aerosol size distribution and CCN number. For the ACT simulation shown in Fig. 3.1, biogenic SOA (from monoterpenes and isoprene) increases annual mean CCN concentrations by up to 100 % over the Amazon, whilst reducing annual mean CCN concentrations by up to 10 % over some tropical oceans.

The largest increases in absolute CCN concentration are simulated in regions coincident with substantial primary particle emissions (particularly regions of tropical biomass burning) suggesting an important interaction between SOA and primary particles. In GLOMAP, primary carbonaceous aerosol from fossil fuel combustion and wildfire is initially non-hydrophilic, being emitted into the Aitken insoluble mode. Condensation of soluble gas phase species moves these particles into the hydrophilic Aitken mode, where they are able to act as CCN. Without this physical ageing by SOA, the global annual mean increase in CCN concentration is reduced to 3.6 % (*ACT_mi_noSOAage*), compared to 12.8 % in the equivalent standard run (*ACT_mi*). When the physical ageing requires only one monolayer of soluble material (*ACT_mi_fast_age*), the global mean increase in CCN number concentration is reduced to 8.9 % because carbonaceous particles are more

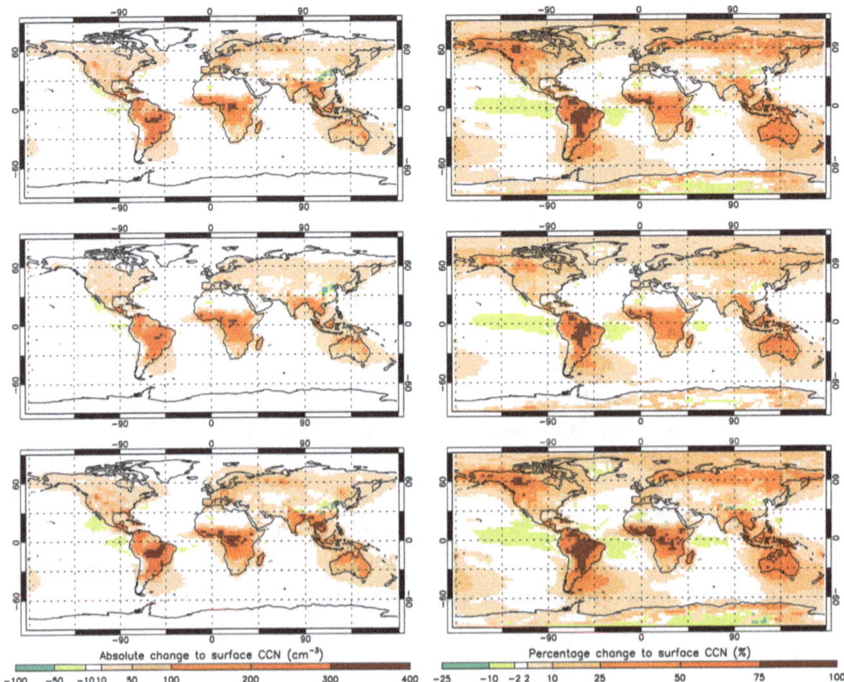


**Fig. 3.1** Simulated annual mean absolute (*left*) and percentage (*right*) changes to surface level cloud condensation nuclei (CCN) number concentration, calculated at 0.2 % supersaturation, resulting from the emission of monoterpenes (*ACT_m*; *upper*), isoprene (*ACT_i*; *middle*) and both monoterpenes and isoprene (*ACT_mi*; *lower*)

efficiently coated and transferred to the hydrophilic distribution by $H_2SO_4$, such that CCN concentrations are less sensitive to the presence of organics. Using a smaller emission size for primary carbonaceous aerosol (*ACT_mi_BCOCsmall*) increases the global annual mean change to surface CCN concentrations when biogenic SOA is included to 17.5 %. A smaller emission size increases the number of primary particles emitted per mass of carbonaceous material, thereby providing more non-hydrophilic particles ready to be aged to the hydrophilic modes where they are able to act as CCN. Additionally, the lower surface area of these smaller particles allows a faster rate of ageing by a given amount of SOA.

At tropical latitudes (30°N to 30°S), high year-round emissions of both isoprene and monoterpenes (Sect. 2.1.2.2) result in large increases to CCN concentrations (Fig. 3.1) throughout the seasonal cycle (red lines in Fig. 3.2, *left*). Between 30 and 90°N, increases in CCN are largest during the NH summer, with small increases simulated in winter months when emissions of BVOC are low. Over the high latitude boreal forests, monoterpene emissions are responsible for the majority of the CCN increase (green lines in Fig. 3.2, *left*), owing to their higher emission rate as compared to isoprene (Sect. 2.1.2.2). In the southern hemisphere

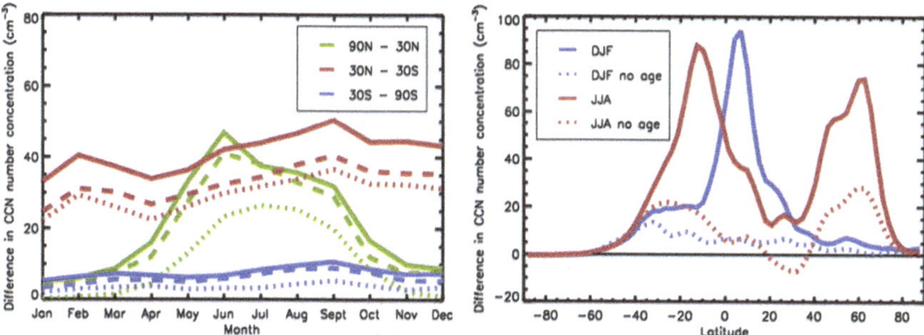


**Fig. 3.2** *Left* Monthly mean absolute change in surface CCN number concentration ($cm^{-3}$) at 0.2 % supersaturation, across three latitude bands, when monoterpene (*ACT_m*; *dashed lines*), isoprene (*ACT_i*; *dotted lines*) and both monoterpene and isoprene (*ACT_mi*; *solid lines*) emissions are included. *Right* Seasonal mean absolute change in CCN number concentration ($cm^{-3}$) at 0.2 % supersaturation during Dec–Feb (*blue*) and Jun–Aug (*red*) using the ACT mechanism; *solid lines* represent *ACT_mi* and *dotted lines ACT_mi_noSOAage* (i.e. condensable organics do not transfer non-hydrophilic particles to the hydrophilic distribution)

(30–90°S), absolute CCN changes are small throughout the year due to low BVOC emissions at these latitudes (blue lines in Fig. 3.2, *left*). However, relatively low absolute changes can result in substantial fractional changes, particularly over ocean regions (Fig. 3.1, *right*) due to low background CCN number concentration.

Within 30° either side of the equator, BVOC emissions in the GEIA inventory [6] are slightly higher during the respective wet seasons (April–September in northern tropics; October–March in southern tropics), but as shown in Fig. 3.2 (*right*), the largest absolute increase in CCN concentration occurs during the dry seasons when primary carbonaceous emissions from wildfires are highest. The importance of ageing is confirmed by much lower CCN increase simulated when secondary organic material does not transfer non-hydrophilic particles to the hydrophilic distribution (dotted lines in Fig. 3.2, *right*). At high northern latitudes, the process of physical ageing also contributes to the summertime CCN increase (Fig. 3.1, *right*), but a more substantial contribution here comes from the growth of smaller particles to CCN active sizes via condensation of organic material.

Over some ocean regions, BVOC emissions can cause reductions in simulated CCN concentrations (Fig. 3.1), as a result of several different processes. The presence of biogenic SOA allows non-hydrophilic particles to be aged (i.e., transferred to a hydrophilic mode) and enhances their growth up to the size at which they may be nucleation scavenged (Sect. 2.1.4.7). In the absence of SOA, simulated particle growth may only proceed via coagulation and condensation of sulphuric acid. Therefore, in the presence of SOA, particles grow more quickly to a size where they may be removed from the atmosphere by nucleation scavenging. Additionally, the presence of biogenic SOA enhances the condensation sink over continental regions, resulting in increased condensation of the $H_2SO_4$ onto existing particles. This can lower $H_2SO_4$ concentrations in the upper troposphere, subsequently reducing binary homogeneous nucleation and the number of particles entrained into the boundary layer. This entrainment of particles formed in the upper troposphere makes the largest contribution to surface CCN concentrations over the sub-tropical oceans [24]. Where these processes outweigh the generation of new CCN via particle growth, and ageing of primary particles, a net reduction in CCN concentration is simulated.

Monoterpene emissions contribute a greater increase to global annual mean CCN concentration, than isoprene emissions (*ACT_m* and *ACT_i*; Table 3.3). This occurs partly because a greater amount of SOA is generated from their oxidation, despite total annual monoterpene emissions in the GEIA inventory being a factor of four lower than those for isoprene, but may also be due to the spatial distribution of emissions and relative proximity to sources of fine particles that require growth to reach CCN sizes (e.g., carbonaceous particles from fossil fuel combustion). As indicated in Figs. 3.1 and 3.2 (*left*), the contributions from each BVOC are not additive; monoterpene and isoprene SOA increase CCN concentrations by approximately 10.4 and 8.5 % respectively whereas the combined emission results in only 12.8 % increase. This suggests a saturation of the global CCN response to the presence of SOA and is confirmed by the reduced sensitivity to yield increase observed for Experiments 4 to 7.

### 3.3.3 Sensitivity to New Particle Formation

The simulated contribution of SOA to global mean CCN concentrations depends strongly on the nucleation mechanism used in the model. Inclusion of an empirically derived particle activation mechanism within the boundary layer (i.e., ACT, Eq. (3.1) results in a greater absolute global annual change in CCN concentration due to monoterpene SOA (+23.5 $cm^{-3}$; *ACT_m*), when compared to the equivalent simulation using only BHN (+20.3 $cm^{-3}$; *BHN_m*). This occurs because particles formed by the activation of $H_2SO_4$ clusters in the boundary layer are able to grow to CCN active sizes by the condensation of organic oxidation products and is particularly evident between 40 and 60°N (Fig. 3.3, *left*) where there is a strong contribution to total particle numbers from nucleation within the boundary layer [24]. However, a similar annual global mean fractional CCN change (+10.4 %) due to monoterpene SOA is simulated in each case, owing to the higher background CCN concentration when the ACT mechanism is used.

When monoterpene oxidation products are allowed to participate directly in nucleation (*Org1_m*, *Org2_m* and *Org3_m*), the contribution of biogenic SOA to CCN concentrations is substantially greater; increasing the global annual mean by 15.5 % for Org2, 35.8 % for Org3 and 45.2 % for Org1. The large increase in CCN concentrations when the Org1 and Org3 mechanisms are used can be attributed to the fact that in the absence of BVOC emissions, new particle formation in the boundary layer does not occur. With these mechanisms, the peak annual mean absolute increase in CCN concentration occurs at approximately 40°N (Fig. 3.3, *left*) due to large CCN increases simulated over south east U.S.A,

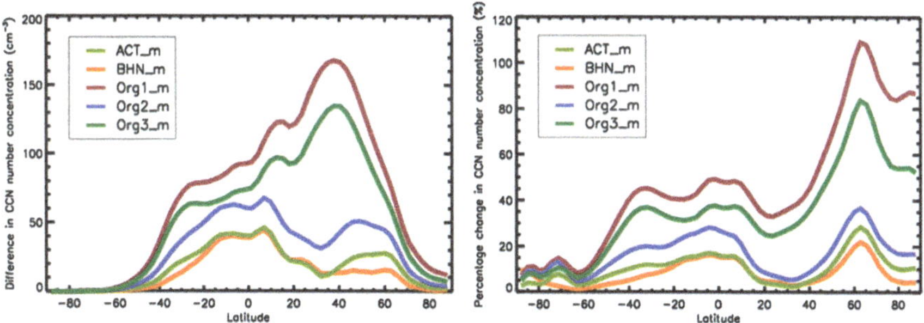


**Fig. 3.3** Annual mean absolute (*left*) and percentage (*right*) change in surface CCN number concentration, at 0.2 % supersaturation, when monoterpene emissions are included in the model, using four particle formation mechanisms: ACT (*light green*), BHN only (*orange*), Org1 (*red*), Org2 (*blue*) and Org3 (*dark green*)

Europe and China, regions of high monoterpene emission during the Northern Hemisphere summer. Substantial fractional increases (over 80 %) are simulated in regions where high monoterpene emissions combine with low background aerosol number concentrations such as the boreal regions of northern Russia and Canada (Figs. 3.3, *right* and 3.4, *right*).

## 3.4 Comparison to Observations

### *3.4.1 Seasonal Cycle at Forested Sites*

The simulated seasonal cycle in the number concentration of particles with dry diameter greater than 80 nm ($N_{80}$) was compared against multi-annual monthly mean observations at three forested sites: Hyytiälä, Finland (e.g. [20, 19]) from 1996–2006, Pallas, Finland (e.g. [7, 15, 16]) from 2000–2011, and Aspvreten, Sweden (e.g. [40]) from 2005–2007. These locations were chosen since they are relatively remote from anthropogenic aerosol sources and are in regions with substantial BVOC emissions. $N_{80}$ concentrations were evaluated since these match particles of CCN relevant size; although $N_{80}$ does not take into account the composition or solubility of the particles, long-term observations of CCN are not yet available at locations suitable for this study. Observations were taken from the EBAS database (available at http://ebas.nilu.no) and monthly mean model values are linearly interpolated to each location.

Table 3.4 gives the Pearson correlation coefficient ($R$) between simulated and observed monthly mean $N_{80}$ concentrations at each site; possible values of R span from -1 (perfect anti-correlation) to +1 (perfect correlation). Figure 3.5 shows the observed and simulated seasonal cycle at Hyytiälä and Pallas; a pronounced seasonal cycle in $N_{80}$ is observed at these locations, with summertime (JJA) concentrations a factor of two greater than those measured in the wintertime (DJF). Without

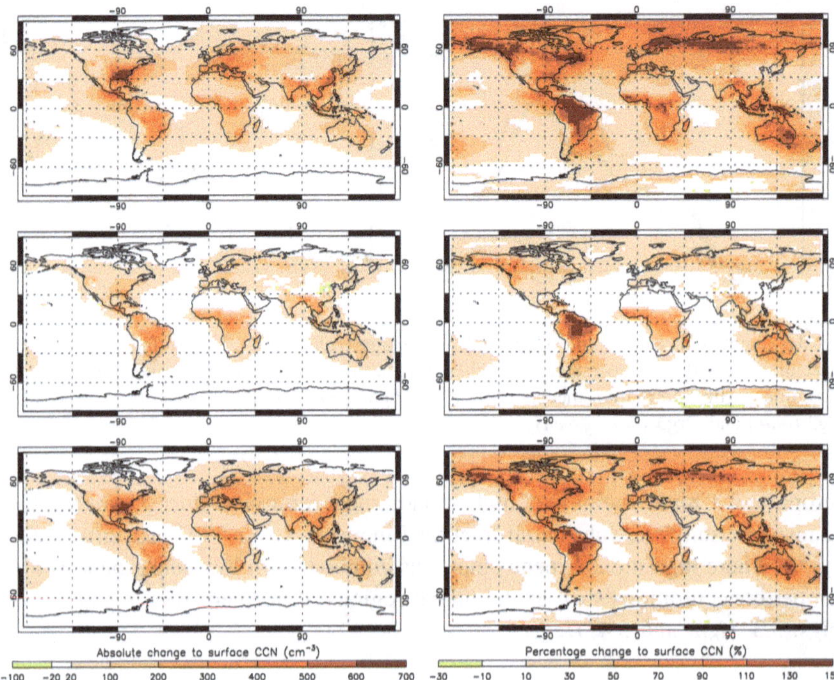


**Fig. 3.4** Simulated annual mean absolute (*left*) and percentage (*right*) changes to surface level cloud condensation nuclei (CCN) number concentration, calculated at 0.2 % supersaturation, when monoterpene emissions are included in the model, for the *Org1_m* (*upper*), *Org2_m* (*middle*) and *Org3_m* (*lower*) experiments. Note that scale differs from Fig. 3.1

biogenic SOA, summertime $N_{80}$ concentrations are under-predicted (dotted line in Fig. 3.5) and simulations do not capture the seasonal cycle; the maximum correlation coefficient at Hyytiälä is 0.37, 0.16 at Aspvreten, and 0.14 at Pallas (all obtained using the Org2 mechanism). The inclusion of biogenic SOA improves the correlation between simulated and observed values for each set of simulations at all three locations, primarily by increasing summertime $N_{80}$ concentrations.

When the ACT mechanism is used, increasing the yield of SOA production (*ACT_mi_x2* and *ACT_mi_x5*) reduces the correlation coefficient at all three sites, when compared to the standard yield simulation (*ACT_mi*). This occurs because, at Hyytiälä for example, summer time $N_{80}$ concentrations simulated with the higher yields are lower than simulated with the standard yield. As indicated in Fig. 3.6 increasing the SOA formation yield increases the size of the largest particles (>400 nm), enhancing the simulated condensation sink for potential nucleating gases and the coagulational sink for nucleation mode particles, suppressing new particle formation and growth as a route to 80 nm particles.

The seasonal cycle is captured best when monoterpene oxidation products are included in the particle formation rate (red, blue and dark green lines in Fig. 3.5),

**Table 3.4** Pearson correlation coefficient (*R*) between multi-annual monthly mean observed and simulated monthly mean $N_{80}$ concentrations

| Simulation | Pearson correlation coefficient (*R*) | | | NMB (%) against subset of CCN dataset |
|---|---|---|---|---|
| | Hyytiälä, Finland. Boreal forest, influenced by European pollution; typical background for high latitude Europe [24.3°E, 61.9°N] | Aspvreten, Sweden. Boreal forest location, mid-Sweden [17.4°E, 58.8°N] | Pallas, Finland. Very remote location at northern border of boreal zone [24.1°E, 68.0°N] | |
| ACT | 0.31 | 0.13 | 0.13 | −44.4 |
| ACT_mi | 0.44 | 0.22 | 0.40 | −16.0 |
| ACT_mi_x0.5 | 0.42 | 0.22 | 0.36 | −16.4 |
| ACT_mi_x2 | 0.41 | 0.19 | 0.35 | −16.7 |
| ACT_mi_x5 | 0.26 | 0.06 | 0.15 | −17.2 |
| ACT_mi_noSOAage | 0.44 | 0.21 | 0.42 | −40.5 |
| ACT_mi_fast_age | 0.44 | 0.19 | 0.40 | −16.5 |
| ACT_mi_BCOCsmall | 0.50 | 0.33 | 0.35 | +48.2 |
| BHN | 0.09 | −0.15 | −0.04 | −48.7 |
| BHN_m | −0.01 | −0.18 | −0.22 | −22.9 |
| Org1 | 0.09 | −0.15 | −0.04 | −48.6 |
| Org1_m | 0.61 | 0.46 | 0.60 | +16.9 |
| Org2 | 0.37 | 0.16 | 0.14 | −26.6 |
| Org2_m | 0.60 | 0.43 | 0.57 | +20.8 |
| Org3 | 0.09 | −0.15 | −0.04 | −48.6 |
| Org3_m | 0.64 | 0.48 | 0.63 | +0.3 |

The normalised mean bias (NMB; Eq. 3.10 in surface CCN concentration between model and observations is also given for each simulation

with the Org3 mechanism giving the best correlation at all three locations (0.64 at Hyytiälä, 0.48 at Aspvreten and 0.63 at Pallas). Figure 3.7 indicates that organically mediated nucleation is required to simulate sufficient particle concentrations between 20–100 nm; however, the number of particles in the nucleation mode is over-predicted.

### 3.4.2 Seasonal Cycle in Total Particle Concentration

The simulated seasonal cycle in total particle number concentration was compared to the multi-annual observational dataset compiled by [36]. Here, data collected

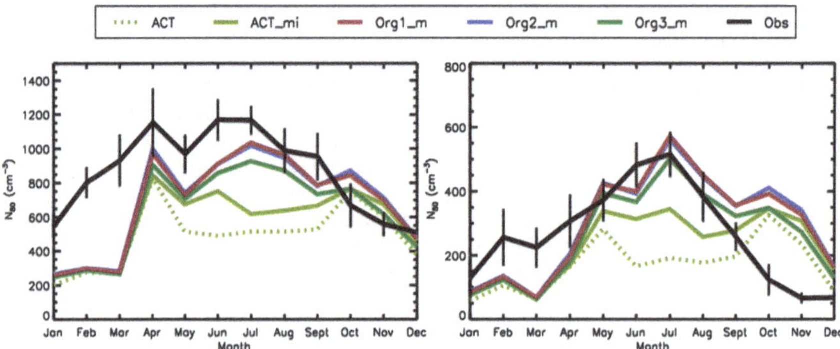


**Fig. 3.5** Multi-annual monthly mean observed (*black lines*) seasonal cycle in $N_{80}$ concentration at Hyytiälä (*left*) and Pallas (*right*); standard deviation of the observed monthly mean is indicated by the vertical black lines. $N_{80}$ concentrations simulated using the ACT (*light green* Expt.1 and Expt. 4) and Org1 (*red*), Org2 (*blue*) and Org3 (*dark green*) nucleation mechanisms

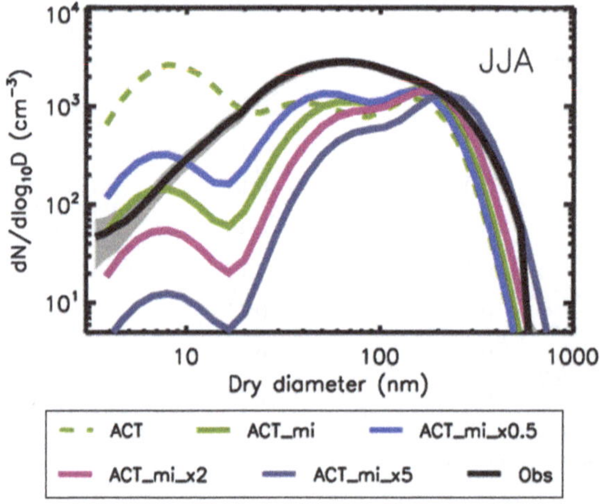


**Fig. 3.6** Simulated and measured (multi-annual; 1996–2006) seasonal (June–July–August) mean number size distribution at Hyytiälä, Finland. Simulations include both monoterpene and isoprene emissions (except *ACT*, *green dashed line*), with SOA yields varied by a factor of 10

from 19 northern hemisphere continental boundary layer locations (Table 3.5) is used. The definition of the particle number concentration varies between sites (e.g., $N_3$, $N_7$ …) due to differences in instrumentation, and the associated minimum cut-off diameter.

A normalised anomaly ($A_x$) for observed monthly mean particle number concentrations ($N_x$), relative to the annual mean particle number concentration ($M_x$), was calculated at each site, as in Eq. 3.9:

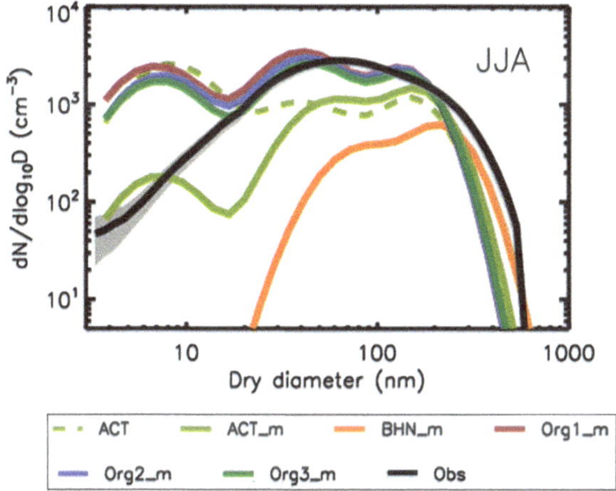


**Fig. 3.7** Simulated and measured (multi-annual; 1996–2006) seasonal (June–July–August) mean number size distribution at Hyytiälä, Finland. Simulations include different representations of new particle formation

**Table 3.5** Observation sites used in comparison; taken from [36]

| Location | | Observation period | Minimum cut-off diameter (nm) |
|---|---|---|---|
| Hyytiälä | 24.3°E, 61.9°N | 2000–2004 | 3 |
| Pallas | 24.1°E, 68.0°N | 2000–2004, 2007 | 10 |
| Finokalia | 25.7°E, 35.3°N | 1997, 2006–2007 | 10 |
| Hohenpeissenberg | 11.0°E, 47.8°N | 2006–2007 | 3 |
| Melpitz | 12.3°E, 51.2°N | 1996–1997, 2003 | 3 |
| Bondville | 88.4°W, 40.1°N | 1994–2007 | 14 |
| Southern great plains | 97.5°W, 36.6°N | 1996–2007 | 10 |
| Tomsk | 85.1°E, 56.5°N | 2005–2006 | 3 |
| Listvyanka | 104.9°E, 51.9°N | 2005–2006 | 3 |
| Harwell | 359.0°E, 51.0°N | 2000 | 10 |
| Weybourne | 1.1°E, 53.0°N | 2005 | 10 |
| India Himilaya | 79.6°E, 29.4°N | 2005–2008 | 10 |
| Aspvreten | 17.4°E, 58.8°N | 2000–2006 | 10 |
| Utö | 21.4°E, 59.8°N | 2003–2006 | 7 |
| Värriö | 29.6°E, 67.8°N | 1998–2006 | 8 |
| Thompson farm | 289.1°E, 43.1°N | 2001–2009 | 7 |
| Castle springs | 71.3°W, 43.7°N | 2001–2008 | 7 |
| Tannus observatory | 8.4°E, 50.2°N | 2008–2009 | 10 |
| Po valley | 11.6°E, 44.7°N | 2002–2006 | 3 |

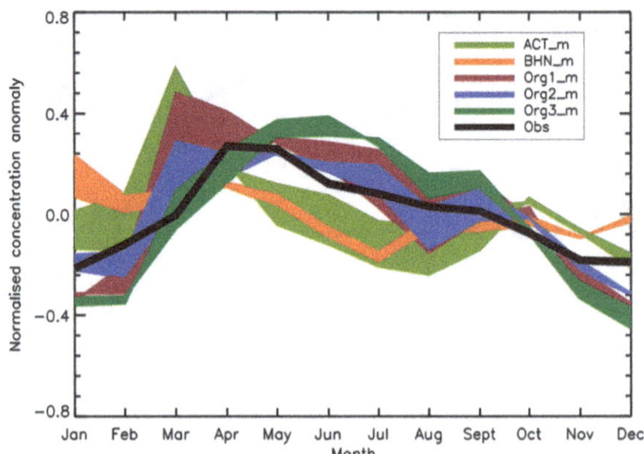


**Fig. 3.8** Normalised concentration anomaly in simulated and observed monthly mean total particle concentration across 19 northern hemisphere continental boundary layer locations [36]. Multi-annual observations, represented by the *black line*, are derived from measurements with various minimum cut-off diameters (Table 3.5); the coloured shading represents the region between the modelled normalised concentration anomaly at 3 and 14 nm

$$A_x = \frac{(N_x - M_x)}{M_x} \tag{3.9}$$

The same approach was applied to the model output from five simulations (*ACT_m*, *BHN_m*, *Org1_m*, *Org2_m* and *Org3_m*); model data were interpolated to the observation location in the horizontal, and in the vertical where necessary. Figure 3.8 compares the observed mean (across the 19 locations) normalised anomaly for each month, to those simulated by the model using five different schemes for new particle formation. Using the ACT particle formation mechanism gives peak particle number concentration during the early spring, with summertime nucleation suppressed by the higher condensation sink. Inclusion of an organically mediated new particle formation mechanism greatly improves the representation of the observed seasonal cycle.

Table 3.6 reports the mean $R$ value, across the 19 locations, between measured and simulated particle concentrations; model values were sampled at the relevant particle size for each location, e.g., to compare against observations at Hyytiälä, $N_3$ was used, but to compare against observations at Pallas, $N_{10}$ was used. Inclusion of an organically mediated new particle formation rate increases $R$ from 0.23 (*ACT_m*; where new particle formation is based on activation of sulphuric acid clusters), up to 0.40 for *Org3_m*.

**Table 3.6** Mean Pearson correlation coefficient between multi-annual observed and simulated monthly mean total particle concentration, across 19 continental boundary layer locations

| Simulation | Pearson correlation coefficient ($R$) |
|---|---|
| ACT_m | 0.23 |
| BHN_m | 0.08 |
| Org1_m | 0.37 |
| Org2_m | 0.33 |
| Org3_m | 0.40 |

**Table 3.7** Locations included in the subset of CCN observations taken from [38]

| Location | | Reference |
|---|---|---|
| Balbina, Amazon Basin | 59.4°W, 1.9°S | [31] |
| Rondonia, Amazon Basin | 61.9°W, 10.9°S | [43] |
| Amazon Basin | 73°W, 5°S and 63°W, 12°S | [1] |
| Fazenda Nossa, Amazon Basin | 62.35°W, 10.8°S | [42] |
| Reno, USA | 119.8°W, 39.5°N | [10] |
| Lauder, New Zealand | 169.7°E, 45°S | [2] |

### 3.4.3 Cloud Condensation Nuclei Concentrations

Simulated CCN concentrations were compared to a subset of the CCN dataset compiled by [38]. The treatment of primary carbonaceous emissions has been shown to strongly influence particle number concentrations and aerosol size distributions simulated by global aerosol microphysics models [28, 38]. Therefore, simulated CCN were compared against measurements filtered to minimise the influence of these particles: that is, data for terrestrial locations with a simulated present-day/pre-industrial CCN concentration ratio (calculated from $[CCN]_{ACT\_mi}$ / $[CCN]_{ACT\_1750\_mi}$) less than 2, during times when the site was reported to be unaffected by wildfire emissions. This subset of data contained 25 observations (each representing time-weighted mean CCN concentration from a sampling period of days to weeks) from the 6 locations detailed in Table 3.7. Relative uncertainties in the observational dataset all lie in the range ±5–40 %, but most within ±10–20 % [38]. CCN concentrations from the model were calculated for each of the six locations using the supersaturation at which the observations were recorded.

Table 3.4 reports the normalised mean bias (NMB) between observed and simulated CCN, calculated according to Eq. 3.10 where $S_x$ are CCN number concentrations simulated by the model, and $O_x$ are observed CCN number concentrations at each location, $x$.

$$NMB = 100\,\% \times \sum (S_x - O_x) \div \sum O_x \tag{3.10}$$

In the absence of biogenic SOA, CCN concentrations are under-predicted (NMB between −48.6 and −26.6 %). Inclusion of biogenic SOA reduces the NMB at these locations to within the uncertainty associated with the observational dataset, e.g., from −44.4 to −16.0 % for *ACT_mi*.

Whilst the mechanisms Org1 and Org2 led to an over prediction in CCN concentration (NMB = +16.9 % for Org1 and +20.0 % for Org2; still within the uncertainty of the measurements), the Org3 mechanism gives very good agreement (NMB = 0.3 %) across this subset of locations. In the Org3 mechanism, *BioOxOrg* is generated only from monoterpene oxidation via OH, rather than the *OxOrg* term in Org1 and Org2 which is generated from monoterpene oxidation by OH, $O_3$ and $NO_3$. This difference in oxidation pathway could introduce spatial, seasonal and diurnal differences to the nucleation rate and warrants further investigation.

When biogenic SOA is included, but is not able to age non-hydrophilic particles to the hydrophilic distribution (*ACT_mi_noSOAage*), CCN concentrations are under-predicted (NMB = −40.5 %) suggesting that despite selecting for relatively pristine locations and times, the ageing of carbonaceous particles still contributes substantially to local CCN concentrations. The faster rate of ageing (*ACT_mi_fast_age*) makes little difference to the correlation (NMB = −16.5 %, as compared to −16.0 % for the standard ageing), suggesting that the process of generating CCN-active particles through physical ageing is not being limited by the availability of condensable material in these locations. This is confirmed by the narrow range of NMB values obtained (−16.0 to −17.2 %) when yield of SOA production is varied by a factor of 10. When the smaller emission size for BC/OC primary particles is used (*ACT_mi_BCOCsmall*), CCN concentrations are substantially over-predicted (NMB = +48.2 %), suggesting that this emission size generates too many CCN active particles in the presence of SOA.

## 3.5 Summary and Conclusions

In all simulations, the inclusion of biogenic SOA increases the global annual mean CCN concentration (by between 3.6 and 45.2 %). In the absence of organic-mediated nucleation, most of the simulated increase in CCN number concentration occurs due to physical ageing, and subsequent growth, of non-hydrophilic particles originating from wildfire and carbonaceous combustion. However, when monoterpene oxidation products affect the new particle formation rate, CCN concentrations are mostly perturbed by the growth of newly formed particles to CCN-active sizes. Similarly, at around 60°N, where monoterpene emissions are high during the northern hemisphere summer months and background particle concentrations are low, a greater proportion of the CCN increase is associated with the growth of smaller particles.

In the absence of organically mediated new particle formation, regional decreases in CCN concentrations are simulated, due to both enhanced nucleation scavenging of non-hydrophilic particles, and the suppression of both upper tropospheric and boundary layer nucleation.

The low sensitivity of CCN to the inclusion of biogenic SOA in experiments without organic-mediated nucleation is consistent with the much larger parameter sensitivity study of [22]. Using an emulator approach, they varied biogenic SOA production between 5 and 360 Tg(SOA) $a^{-1}$, resulting in a global mean 3 % standard deviation in CCN concentration.

Simulated particle concentrations were compared to observations from a range of locations. In the absence of SOA, GLOMAP-mode fails to capture the summertime peak in $N_{80}$ concentrations at boreal forest locations (e.g., $R$ of 0.31 for *ACT* at Hyytiälä), with the best representation found when organic oxidation products contribute directly to new particle formation (e.g., $R$ of 0.64 for *Org3_m* at Hyytiälä). Similarly, the seasonal cycle in total particle concentration at sites across the continental boundary layer is better captured when organically mediated new particle formation is included ($R$ of 0.40 for *Org3_m* as compared to 0.23 for *ACT_m*).

Results in this chapter suggest that organic compounds do contribute to the initial stages of atmospheric new particle formation, and therefore, that CCN sensitivity to the presence of biogenic SOA is high.

## References

1. Andreae MO et al (2004) Smoking rain clouds over the amazon. Science 303(5662):1337–1342
2. Delene DJ, Deshler T (2001) Vertical profiles of cloud condensation nuclei above Wyoming. J Geophys Res Atmos 106(D12):12579–12588
3. Dentener F et al (2006) Emissions of primary aerosol and precursor gases in the years 2000 and 1750 prescribed data-sets for AeroCom. Atmos Chem Phys 6(12):4321–4344
4. Goldstein AH, Galbally IE (2007) Known and unexplored organic constituents in the earth's atmosphere. Environ Sci Technol 41(5):1514–1521
5. Griffin RJ et al (1999) Estimate of global atmospheric organic aerosol from oxidation of biogenic hydrocarbons. Geophys Res Lett 26(17):2721–2724
6. Guenther A et al (1995) A global model of natural volatile organic compound emissions. J Geophys Res 100(D5):8873–8892
7. Hatakka J et al (2003) Overview of the atmospheric research activities and results at Pallas GAW station. Boreal Environ Res 8:365–383
8. Heald CL et al (2010) Satellite observations cap the atmospheric organic aerosol budget. Geophys Res Lett 37(24):L24808
9. Heald CL et al (2011) Exploring the vertical profile of atmospheric organic aerosol: comparing 17 aircraft field campaigns with a global model. Atmos Chem Phys 11(24):12673–12696

10. Hudson JG, Frisbie PR (1991) Surface cloud condensation nuclei and condensation nuclei measurements at Reno, Nevada. Atmos Environ Part A Gen Top 25(10):2285–2299
11. Kanakidou M et al (2005) Organic aerosol and global climate modelling: a review. Atmos Chem Phys 5(4):1053–1123
12. Kanawade VP et al (2011) Isoprene suppression of new particle formation in a mixed deciduous forest. Atmos Chem Phys 11(12):6013–6027
13. Kiendler-Scharr A et al (2009) New particle formation in forests inhibited by isoprene emissions. Nature 461(7262):381–384
14. Kirkby J et al (2011) Role of sulphuric acid, ammonia and galactic cosmic rays in atmospheric aerosol nucleation. Nature 476(7361):429–433
15. Komppula M et al (2003) Observations of new particle formation and size distributions at two different heights and surroundings in subarctic area in northern Finland. J Geophys Res Atmos 108(D9):4295
16. Kroll JH et al (2005) Secondary organic aerosol formation from isoprene photooxidation under high-$NO_x$ conditions. Geophys Res Lett 32(18):L18808
17. Kroll JH et al (2006) Secondary organic aerosol formation from isoprene photooxidation. Environ Sci Technol 40(6):1869–1877
18. Kulmala M et al (1998) Parameterisations for sulphuric acid/water nucleation rates. J Geophys Res Atmos 103(D7):8301–8307
19. Kulmala M et al (1998) Analysis of the growth of nucleation mode particles observed in Boreal forest. Tellus B 50(5):449–462
20. Kulmala M et al (2001) Overview of the international project on biogenic aerosol formation in the boreal forest (BIOFOR). Tellus B 53(4):324–343
21. Kulmala M et al (2006) Cluster activation theory as an explanation of the linear dependence between formation rate of 3 nm particles and sulphuric acid concentration. Atmos Chem Phys 6:787–793
22. Lee LA et al (2013) The magnitude and causes of uncertainty in global model simulations of cloud condensation nuclei. Atmos Chem Phys 13:8879–8914
23. Mann GW et al (2012) Intercomparison of modal and sectional aerosol microphysics representations within the same 3-D global chemical transport model. Atmos Chem Phys 12(10):4449–4476
24. Merikanto J et al (2009) Impact of nucleation on global CCN. Atmos Chem Phys 9:8601–8616
25. Metzger A et al (2010) Evidence for the role of organics in aerosol particle formation under atmospheric conditions. Proc Natl Acad Sci 107(15):6646–6651
26. Paasonen P et al (2010) On the roles of sulphuric acid and low-volatility organic vapours in the initial steps of atmospheric new particle formation. Atmos Chem Phys 10(22):11223–11242
27. Petters MD et al (2006) Chemical aging and the hydrophobic-to-hydrophilic conversion of carbonaceous aerosol. Geophys Res Lett 33(24):L24806
28. Reddington CL et al (2011) Primary versus secondary contributions to particle number concentrations in the European boundary layer. Atmos Chem Phys 11(23):12007–12036
29. Riccobono F et al (2014) Oxidation products of biogenic emissions contribute to nucleation of atmospheric particles. Science 344(6185):717–721
30. Riipinen I et al (2011) Organic condensation: a vital link connecting aerosol formation to cloud condensation nuclei (CCN) concentrations. Atmos Chem Phys 11(8):3865–3878
31. Roberts GC et al (2001) Cloud condensation nuclei in the Amazon Basin: "marine" conditions over a continent? Geophys Res Lett 28(14):2807–2810
32. Spracklen DV et al (2005) A global off-line model of size-resolved aerosol microphysics: II. Identification of key uncertainties. Atmos Chem Phys 5(12):3233–3250
33. Spracklen DV (2005) Development and application of a global model of aerosol processes, University of Leeds. UK, PhD
34. Spracklen DV et al (2006) The contribution of boundary layer nucleation events to total particle concentrations on regional and global scales. Atmos Chem Phys 6(12):5631–5648

35. Spracklen DV et al (2008) Contribution of particle formation to global cloud condensation nuclei concentrations. Geophys Res Lett 35(6):L06808
36. Spracklen DV et al (2010) Explaining global surface aerosol number concentrations in terms of primary emissions and particle formation. Atmos Chem Phys 10(10):4775–4793
37. Spracklen DV et al (2011) Aerosol mass spectrometer constraint on the global secondary organic aerosol budget. Atmos Chem Phys 11(23):12109–12136
38. Spracklen DV et al (2011) Global cloud condensation nuclei influenced by carbonaceous combustion aerosol. Atmos Chem Phys 11(17):9067–9087
39. Stier P et al (2005) The aerosol-climate model ECHAM5-HAM. Atmos Chem Phys 5(4):1125–1156
40. Tunved P et al (2004) An investigation of processes controlling the evolution of the boundary layer aerosol size distribution properties at the Swedish background station Aspvreten. Atmos Chem Phys 4(11/12):2581–2592
41. Tunved P et al (2004) A pseudo-Lagrangian model study of the size distribution properties over Scandinavia: transport from Aspvreten to Värriö. Atmos. Chem. Phys. Discuss. 4(6):7757–7794
42. Vestin A et al (2007) Cloud-nucleating properties of the Amazonian biomass burning aerosol: cloud condensation nuclei measurements and modeling. J Geophys Res Atmos 112(D14):D14201
43. Williams E et al (2002) Contrasting convective regimes over the Amazon: implications for cloud electrification. J Geophys Res Atmos 107(D20):8082

# Chapter 4
# The Radiative Impact of Biogenic SOA

## 4.1 Introduction

Despite the ubiquity of organic material in the particle phase over much of the world (e.g., [12, 30]), its impact on the climate is not well understood [17].

In Chap. 3, the presence of biogenic SOA was shown to alter the number and size of particles in the atmosphere; specifically increasing the global annual mean concentration of CCN-sized particles. In this chapter, the direct and first aerosol indirect radiative effects resulting from these changes to the aerosol distribution are quantified using the Edwards-Slingo radiative transfer model.

The top-of-atmosphere (TOA) change in radiative flux due to the presence of biogenic SOA is referred to in this thesis as its radiative effect (RE), after Rap et al. [24]. Also calculated in this chapter is the change in TOA radiative balance relative to pre-industrial conditions, termed the radiative forcing (RF), after Forster et al. [8] and Ramaswamy et al. [23].

## 4.2 Experimental Setup

### *4.2.1 The Edwards–Slingo Radiative Transfer Model*

The radiative transfer model of Edwards and Slingo [7] (hereafter E–S) calculates radiative fluxes, or radiances, given a particular atmospheric state. The E–S model was originally designed to operate within the Hadley Centre Global Environmental Model (HadGEM) but is used here in an offline configuration, with a horizontal resolution of 2.5° × 2.5° and 23 fixed vertical levels (from the surface to 1 hPa). To avoid the computational expense involved in modelling all atmospheric absorption lines, the spectrum of radiation is split into six bands in the shortwave (SW) region, and nine bands in the longwave (LW) region, with each band modelled as a downward and upward flux.



C.E. Scott, *The Biogeochemical Impacts of Forests and the Implications for Climate Change Mitigation*, Springer Theses, DOI 10.1007/978-3-319-07851-9_4

In the offline configuration, multi-annual mean (1990–2000) monthly climatologies for gas-phase species (e.g., water vapour, $O_3$, $CO_2$, $CH_4$) and temperature are used, based on ECMWF reanalysis data. Cloud fields (liquid water path and cloud fraction) are taken from the ISCCP-D2 archive [25] for the year 2000. Sensitivity of the direct and first indirect radiative effects to the particular cloud climatology used (i.e., single year versus multi-annual mean) was examined in Rap et al. [24] and found be to be very low, for a range of natural aerosol sources including monoterpene derived SOA.

### *4.2.2 Direct Radiative Effect*

To determine the direct radiative effect (DRE), the radiative transfer model was used to calculate the difference in net TOA all-sky (i.e., including clouds) radiative flux (SW + LW) between experiments including SOA and the equivalent experiments without SOA.

The offline E–S model was adapted to include a new offline version of the UKCA_RADAER code, previously configured for use with the CLASSIC (Coupled Large-scale Aerosol Simulator for Studies In Climate) aerosol scheme (e.g., [4]). However, because the mean radius of, chemical composition, and amount of water associated with each aerosol mode varies during simulations with GLOMAP-mode, the aerosol optical properties cannot be pre-computed, as was the case with previous implementations of UKCA_RADAER [3, 4]. Accordingly, the modal refractive index is calculated as the volume-weighted average refractive index of the individual components (given in Table A.1 of Bellouin et al. [4]) in the mode, including water. As described by Bellouin et al. [3], optical properties are obtained from a look-up table of all possible combinations of refractive index and Mie parameter (relationship between the modal radius and the wavelength of radiation) for computational efficiency.

In the GLOMAP experiments described in Chap. 3, secondary organic material is added to the POM component. When coupling to UKCA_RADAER, the optical properties of POM are calculated assuming the characteristics of *aged fossil fuel organic carbon* (Table A.1; Bellouin et al. [4]) to reflect the fact that POM contains a mixture of primary and secondary organic components.

The direct radiative effect calculations were performed by Alexandru Rap, but the candidate performed all subsequent analysis.

### *4.2.3 Aerosol Indirect Effect*

The first aerosol indirect effect (AIE), or cloud albedo effect, describes the radiative perturbation resulting from a change to the number concentration of cloud droplets, when a fixed cloud water content is assumed [16, 23, 29]. As such, an accurate estimation of the cloud droplet number concentration (CDNC), before and after any given perturbation, is required.

#### 4.2.3.1 Determining Cloud Droplet Number Concentrations

Empirical relationships are often used to determine CDNC from aerosol mass or number concentration (e.g., [5, 13]). However, these relationships do not account for the distribution of particle sizes and composition, or the maximum supersaturation of individual air parcels [22]. As an alternative, mechanistic parameterisations may be used to determine CDNC as a function of precursor aerosol concentrations and updraught velocity (e.g. [1, 9, 19]).

In the work presented here, CDNC are calculated using the parameterisation developed by Nenes and Seinfeld [19], as updated by Fountoukis and Nenes [9] and Barahona et al. [2]. Merikanto et al. [18] previously demonstrated good agreement between aircraft measurements and CDNC values derived using this approach in combination with the GLOMAP model. The monthly mean aerosol size distribution is converted to a supersaturation distribution from which the number of activated particles can be determined at a given supersaturation. The maximum supersaturation ($SS_{max}$) is computed for an adiabatic parcel and occurs when water availability from parcel cooling becomes equal to the rate at which water vapour is depleted by condensation onto activated particles. As such, $SS_{max}$ depends upon the aerosol population and must be diagnosed at each point in time and space, rather than prescribed uniformly; in polluted environments with abundant potential CCN, there will be many sites for the condensation of water, and $SS_{max}$ will be suppressed.

In this chapter, CDNC are calculated with a uniform updraught speed of 0.15 m $s^{-1}$ over sea and 0.3 m $s^{-1}$ over land, in line with those commonly observed for stratus clouds [11, 21]. The global distribution of updraught velocities is uncertain, so the sensitivity of the radiative impact of biogenic SOA to the choice of updraught velocity is examined in Chap. 5.

#### 4.2.3.2 Determining the AIE

The effective radius ($r_e$) of a cloud droplet may be expressed relative to the density of water ($\rho_w$; g $cm^{-3}$), and the liquid water path ($LWP$; g $m^{-2}$), thickness ($\Delta z$; m) and CDNC ($cm^{-3}$) of the cloud, as in Eq. 4.1 [6]:

$$r_e = 100 \times \left[\frac{LWP}{\Delta z} \times \frac{3}{4\pi\rho_w CDNC}\right]^{\frac{1}{3}} \tag{4.1}$$

For each perturbation experiment, the AIE is calculated relative to an equivalent control experiment. A uniform control cloud droplet effective radius ($r_{e1}$) of 10 μm is assumed to maintain consistency with the approach used by ISCCP to derive the *LWP*. For each perturbation experiment, a new field of effective radii ($r_{e2}$) is calculated using Eq. 4.2 (obtained by assuming fixed *LWP* and $\Delta z$, between experiments, in Eq. 4.1; Spracklen et al. [28]); $CDNC_1$ represents the simulation including SOA, and $CDNC_2$ represents the simulation with no SOA. The effective radii are modified only for low- and mid-level clouds, up to 600 hPa.

$$r_{e2} = r_{e1}\left[\frac{CDNC_1}{CDNC_2}\right]^{\frac{1}{3}} \tag{4.2}$$

The first AIE of biogenic SOA is then calculated by comparing net (SW + LW) radiative fluxes using the $r_{e2}$ values derived for each perturbation experiment, to those of the control simulation with fixed $r_{e1}$. In these offline experiments, the second aerosol indirect (cloud lifetime) effect is not calculated.

## 4.3 Results

### *4.3.1 Direct Radiative Effect*

An annual global mean DRE of between −0.08 and −0.78 W $m^{-2}$ is calculated for biogenic SOA in the present-day atmosphere (Table 4.1), with the spatial distribution of the effect for *ACT_mi* shown in Fig. 4.1 (left). The DRE from biogenic SOA is strongest at tropical latitudes, with local annual mean DREs as large as −4 W $m^{-2}$ over the Amazon and central Africa.

Figure 4.1 (right) shows the sensitivity of the global annual mean DRE to the processes examined in Chap. 3. The magnitude of the direct effect is highly sensitive to the amount of SOA (and therefore the amount of potential particle growth) included in the simulation, with the global annual mean varying from −0.09 W $m^{-2}$ when the SOA production yield is halved (source of 18.5 Tg(SOA) $a^{-1}$), to −0.78 W $m^{-2}$ when the yield is increased by a factor of 5 (source of 185.1 Tg(SOA) $a^{-1}$).

Varying the nucleation mechanism, results in little variability in the size of larger particles (e.g. Fig. 3.8), and therefore little variability in the DRE (range −0.08 to −0.10 W $m^{-2}$). The large increase in the number of nucleated particles and CCN simulated with the organically mediated nucleation mechanisms is less important for the DRE as these particles are too small to influence the path of incoming solar radiation.

The DREs simulated by Rap et al. [24] and O' Donnell et al. [20] lie within the range we calculate here (Table 4.2). Goto et al. [10] calculated a smaller DRE of −0.01 W $m^{-2}$ but used a mass-only aerosol model which does not simulate the growth of particles associated with the condensation of secondary organic material.

**Table 4.1** Global annual mean changes to CDNC at cloud height (approximately 900 hPa), all-sky DRE and first AIE, relative to an equivalent control simulation including no BVOC emission

| Experiment No. | Experiment name | ΔCDNC ($cm^{-3}$) | All-sky DRE ($W\ m^{-2}$) | First AIE ($W\ m^{-2}$) |
|---|---|---|---|---|
| 2 | ACT_m | +7.7 (+4.1 %) | −0.10 | −0.07 |
| 3 | ACT_i | +5.9 (+3.2 %) | −0.08 | −0.06 |
| 4 | ACT_mi | +8.2 (+4.4 %) | −0.18 | −0.06 |
| 5 | ACT_mi_x0.5 | +7.1 (+3.8 %) | −0.09 | −0.07 |
| 6 | ACT_mi_x2 | +9.0 (+4.8 %) | −0.33 | −0.04 |
| 7 | ACT_mi_x5 | +9.3 (+5.0 %) | −0.78 | +0.01 |
| 8 | ACT_mi_noSOAage | +3.6 (+1.9 %) | −0.14 | −0.02 |
| 10 | ACT_mi_fast_age | +4.9 (+2.5 %) | −0.18 | −0.02 |
| 12 | ACT_mi_BCOCsmall | +10.4 (+5.2 %) | −0.18 | −0.12 |
| 14 | BHN_m | +5.8 (+3.5 %) | −0.10 | −0.05 |
| 16 | Org1_m | +44.2 (+26.6 %) | −0.08 | −0.77 |
| 18 | Org2_m | +14.3 (+7.2 %) | −0.10 | −0.22 |
| 20 | Org3_m | +34.3 (+20.7 %) | – | −0.59 |

## *4.3.2 First Aerosol Indirect Radiative Effect*

### 4.3.2.1 Cloud Droplet Number Concentration

The inclusion of biogenic SOA increases the calculated global annual mean CDNC by between 1.9 and 5.2 % when the ACT nucleation mechanism is used (Table 4.1). Figure 4.2 shows the spatial distribution of the annual mean change in CDNC due to the presence of biogenic SOA in the *ACT_mi* simulation. Over most regions, perturbations to calculated CDNC follow the same spatial pattern as the changes to CCN concentration described in Chap. 3. However, in regions with high pre-existing aerosol concentrations, such as those heavily affected by biomass burning, activation of additional CCN to cloud droplets can become limited by competition for water vapour. This is evident over South America and western Africa, where the largest changes to CDNC (Fig. 4.2) do not coincide spatially with the largest changes to CCN (Fig. 3.2, lower).

As with CCN, relatively low absolute changes to CDNC in the southern hemisphere can lead to high fractional changes over the oceans. In the boreal regions, the inclusion of biogenic SOA increases the annual mean CDNC by up to 70 % (Fig. 4.2, right), due to very low background CDNC and therefore a high sensitivity to additional CCN. Small decreases (<10 %) in CDNC occur over some ocean regions due to the decreases in CCN described in Chap. 3.

Including monoterpene oxidation products in the particle formation rate equation yields the greatest increase in CDNC due to biogenic SOA (e.g., +20.7 %

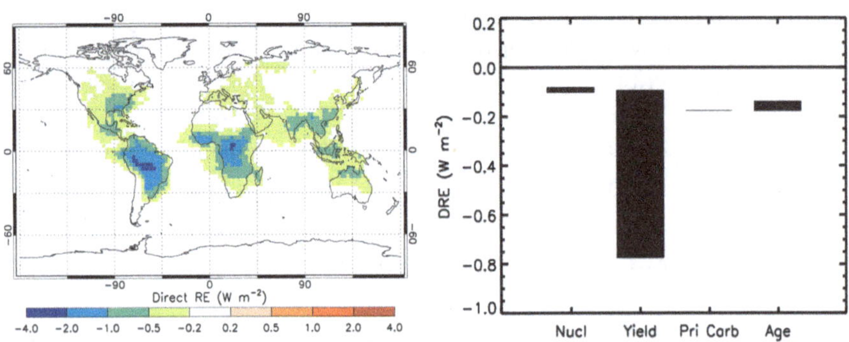


**Fig. 4.1** Annual mean all-sky DRE (*left*) due to biogenic SOA (*ACT_mi*) relative to an equivalent simulation with no biogenic SOA. Variation in the global annual mean DRE (*right*) of biogenic SOA associated with several parameters; black bars indicate the range of values obtained for each set of experiments: *Nucl* (nucleation mechanism; Expt. 2, 14, 16, 18, 20), *Yield* (SOA yield; Expt. 4, 5, 6, 7), *Pri Carb* (primary carbonaceous emission size; Expt. 4, 12), *Age* (physical ageing; Expt. 4, 8, 10)

**Table 4.2** Summary of previous estimates of the radiative effects of SOA

| Study | SOA included | Clear-sky DRE (W m$^{-2}$) | All-sky DRE (W m$^{-2}$) | AIE (W m$^{-2}$) |
|---|---|---|---|---|
| Goto et al. [10] | Monoterpenes | not calculated | −0.01 | −0.19 |
| O'Donnell et al. [20] | Monoterpenes and isoprene | −0.29 | not calculated | +0.23[a] |
| Rap et al. [24] | Monoterpenes | −0.14 | −0.13 | −0.02 |
| This study | Monoterpenes/ isoprene/both | not calculated | −0.08 to −0.78 | +0.01 to −0.77 |

[a] Also includes anthropogenic SOA

for Org3). Figure 4.3 shows the spatial distribution of the annual mean CDNC change when the Org3 mechanism is used; as with CCN (Fig. 3.5), large absolute increases are simulated at around 40°N, whilst fractional increases in excess of 100 % are simulated above 50°N (Fig. 4.3).

#### 4.3.2.2 Aerosol Indirect Effect

In the present-day atmosphere, biogenic SOA exerts a global annual mean AIE of between +0.01 and −0.77 W m$^{-2}$ (Table 4.1). Figure 4.4 (left) shows the spatial distribution in the first AIE due to biogenic SOA for the *ACT_mi* simulation. A negative first AIE occurs in locations experiencing a large relative increase in CDNC (e.g., boreal Asia), or a modest increase in CDNC coinciding with a high fraction of low level cloud cover (e.g., Southern Hemisphere oceans). A positive

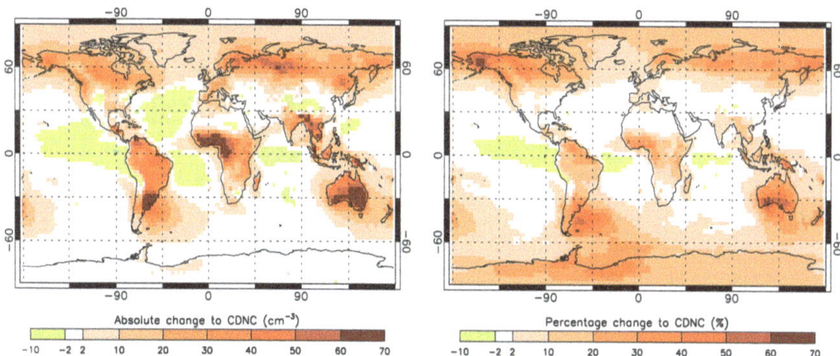


**Fig. 4.2** Annual mean absolute (*left*) and percentage (*right*) change to CDNC due to biogenic SOA for the *ACT_mi* experiment (i.e., $[\mathrm{CDNC}]_{\mathrm{ACT\_mi}}-[\mathrm{CDNC}]_{\mathrm{ACT}}$)

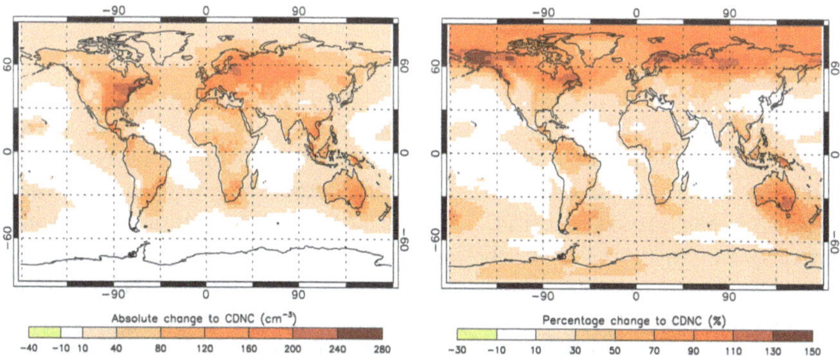


**Fig. 4.3** Annual mean absolute (*left*) and percentage (*right*) change to CDNC due to biogenic SOA for the *Org3_m* experiment (i.e., $[\mathrm{CDNC}]_{\mathrm{Org3\_m}}-[\mathrm{CDNC}]_{\mathrm{Org3}}$)

AIE occurs in locations where a small decrease in CDNC (Fig. 4.2) coincides with very high cloud fractions, such as the western coasts of Central America and the South Atlantic (Fig. 4.4, left).

Figure 4.4 (right) summarises the sensitivity of the global annual mean first AIE to the processes examined in Chap. 3. The simulated global annual mean first AIE is most sensitive to the nucleation mechanism used in the model (global annual mean range −0.05 to −0.77 W $m^{-2}$ for *BHN_m* and *Org1_m*, respectively); the inclusion of monoterpene oxidation products in the particle formation rate results in greater CCN and CDNC increases, and therefore a more negative global annual mean AIE, than either the BHN or ACT mechanisms. As with the ACT simulation, the organic nucleation mechanisms generate a negative AIE over the oceans between 30°S and 50°S (Fig. 4.5), but Org1 and Org3 in particular also result in large negative forcings in the northern hemisphere due to high fractional CCN (Fig. 3.4) and CDNC changes above 40°N (Fig. 4.3).

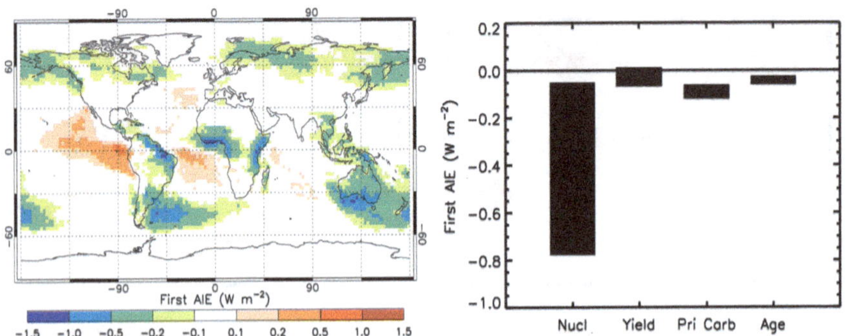


**Fig. 4.4** Annual mean first AIE (*left*) associated with the perturbation in cloud droplet number concentration due to biogenic SOA (*ACT_mi*), relative to an equivalent simulation with no biogenic SOA. Variation in the global annual mean AIE (*right*) of biogenic SOA associated with several parameters; black bars indicate the range of values obtained for each set of experiments: *Nucl* (nucleation mechanism; Expt. 2, 14, 16, 18), *Yield* (SOA yield; Expt. 4, 5, 6, 7), *Pri Carb* (primary carbonaceous emission size; Expt. 4, 12), *Age* (physical ageing; Expt. 4, 8, 10)

Using smaller size characteristics for primary carbonaceous emissions (*ACT_mi_BCOCsmall*) gives greater CCN and CDNC increases due to biogenic SOA, and subsequently a more substantial AIE of $-0.12$ W m$^{-2}$. However, results from Sect. 3.4.3 suggest that this emission size generates too many CCN-sized particles. When secondary organic material is not able to age non-hydrophilic particles to the hydrophilic distribution (*ACT_mi_noSOAage*), the global annual mean AIE reduces to $-0.02$ W m$^{-2}$, due to smaller increases in CCN and CDNC.

Whilst increasing the yield of SOA production gives a greater global annual mean fractional increase in CDNC (Table 4.1), the increased yield enhances the small CDNC decreases in regions with very high cloud fraction, enhancing regions of positive AIE at tropical latitudes (Fig. 4.6). As discussed in Sect. 3.4.1, increasing the SOA production yield also increases the size of the largest particles and supresses the formation and growth of new particles (e.g., Fig. 3.7). Consequently, the dominant change to the size distribution, when biogenic SOA is included at an enhanced yield, but in the absence of organically mediated nucleation, is to increase the size of particles already large enough to act as CCN. Since more water is then required to activate each particle, this results in a smaller increase to CDNC. This process is particularly evident during the northern hemisphere summertime; consequently a smaller global annual mean first AIE is simulated with increasing yield (Table 4.1 and Fig. 4.6).

Over boreal forests, regional annual mean AIEs of between $-0.1$ and $-0.5$ W m$^{-2}$ are calculated for the *ACT_mi* experiment (Fig. 4.4, *left*). When an organic nucleation mechanism is used, e.g., Fig. 4.5 for *Org3_m*, regional annual mean AIEs up to $-2$ W m$^{-2}$ are calculated. As illustrated in Fig. 4.7, much of the boreal region experiences a summertime (JJA mean) first AIE of between $-1$ and $-5$ W m$^{-2}$, when the Org3 mechanism is used, matching the large cooling effect over these forest

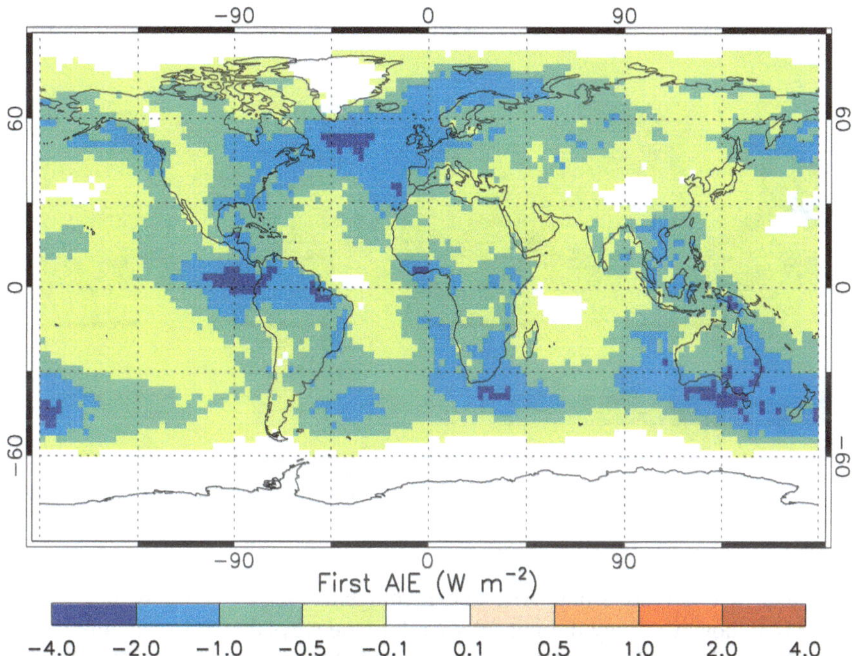


**Fig. 4.5** Annual mean first AIE associated with the perturbation in cloud droplet number concentration due to biogenic SOA in the *Org3_m* simulation

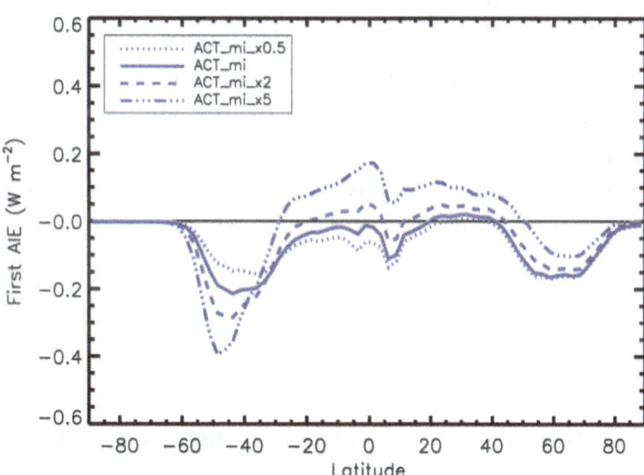


**Fig. 4.6** Annual zonal mean AIE from biogenic SOA when the yield of SOA production is varied by a factor of 10, using the ACT nucleation mechanism

regions calculated by previous studies [14, 27]. However, the strongest radiative effects (up to $-8$ W m$^{-2}$) are simulated over the ocean regions above 50°N which whilst experiencing smaller increases in CDNC, have higher cloud coverage (i.e., cloud fraction of 50–70 % as compared to 0–30 % over the land).

The AIE from monoterpene SOA ($-0.19$ W m$^{-2}$) estimated by Goto et al. [10] lies within the range of AIE we calculate here (Table 4.2). Rap et al. [24]

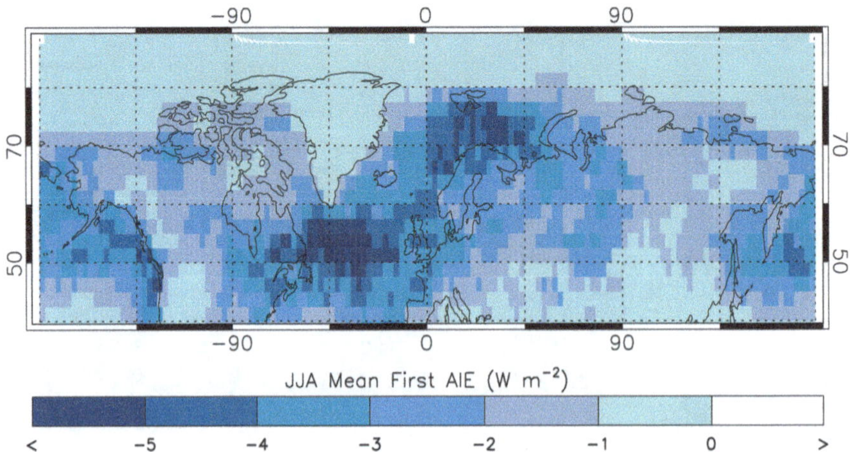


**Fig. 4.7** Summertime (June–July–August) mean first AIE from biogenic SOA in the *Org3_m* experiment

calculated a relatively small AIE due to biogenic SOA (−0.02 W $m^{-2}$) using GLOMAP, however, they did not fully explore the ways that SOA can affect CCN and CDNC. The AIE calculated by Rap et al. [24] was for a GLOMAP-mode simulation using only binary homogenous nucleation (as in *BHN_m*) and requiring only one soluble monolayer to transfer non-hydrophilic particles to the hydrophilic distribution (as in *ACT_mi_fast_age*), and is consistent with the AIEs calculated here (Table 4.1).

O'Donnell et al. [20] simulated a positive AIE (+0.23 W $m^{-2}$) for all SOA (i.e., biogenic plus anthropogenic), which lies outside the range calculated here. This positive AIE may be caused by the approach used by O'Donnell et al. [20] to distribute SOA amongst the existing aerosol size distribution which results in SOA being distributed preferentially amongst larger size particles (i.e., those already large enough to act as CCN). Sensitivity of the AIE from biogenic SOA to assumptions concerning the distribution of SOA across the existing aerosol size distribution is examined in Chap. 5.

## 4.4 Sensitivity to Anthropogenic Emissions

Primary particulate (POM and BC) and gas-phase ($SO_2$) emissions from anthropogenic sources were much lower in 1750 compared to the present day. These lower emissions result in lower simulated background (i.e., in the absence of biogenic SOA) concentrations of CCN and CDN. The impact of biogenic SOA may therefore have been different in a pre-industrial atmosphere. To explore this possibility the impact of biogenic SOA in an atmosphere with 1750 anthropogenic

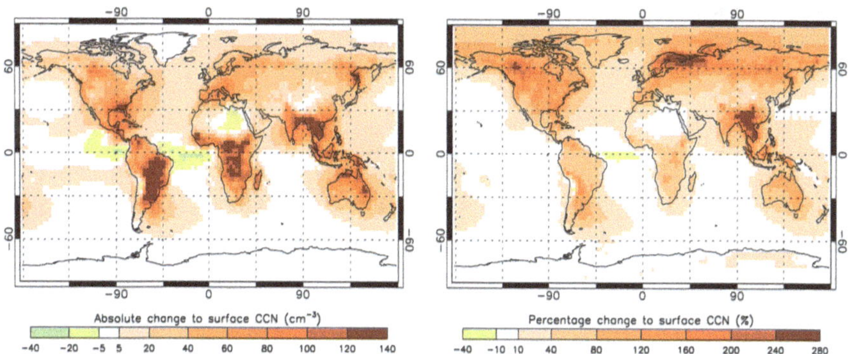


**Fig. 4.8** Simulated annual mean absolute (*left*) and percentage (*right*) changes to surface level CCN number concentration (calculated at 0.2 % supersaturation) due to the inclusion of monoterpene and isoprene emissions in the *ACT_1750_mi* experiment

emissions was examined. In order to isolate the influence of anthropogenic emissions, BVOC emissions were held fixed, but their spatial distribution and magnitude would have been different in the pre-industrial period (e.g., [15]).

Including biogenic SOA in a pre-industrial atmosphere yields a lower absolute change in CCN number concentration (*ACT_1750_mi*; +23.8 $cm^{-3}$) than in the present day (*ACT_mi*; +29.0 $cm^{-3}$). This lower absolute change is due to fewer ultrafine particles being available (from nucleation and primary sources) for growth to CCN sizes, or physical ageing, by the secondary organic material. However, despite the lower absolute changes (Fig. 4.8, left), the inclusion of biogenic SOA results in higher fractional changes (Fig. 4.8, right) in CCN number concentration due to a lower background in the pre-industrial atmosphere.

This increased fractional change in CCN, combined with lower background CDNC in the pre-industrial atmosphere results in a greater perturbation to global annual mean CDNC due to biogenic SOA (*ACT_1750_mi*; +12.6 %), as compared to the present day (*ACT_mi*; +4.4 %). Since the fractional change in CDNC constrains the AIE (Eq. 4.2), a more substantial indirect effect of −0.19 W $m^{-2}$ is simulated with 1750 anthropogenic emissions (Table 4.3) compared to −0.06 W $m^{-2}$ with the same model setup in the present day.

Regions of both positive and negative change in CDNC are enhanced by the lower background concentration, so this global mean radiative effect represents a combination of northern hemisphere land and southern hemisphere ocean regions experiencing a more negative AIE and tropical oceans experiencing a more substantial positive AIE (Fig. 4.9, right), as compared to an equivalent simulation using present-day anthropogenic emissions (Fig. 4.10).

The global mean first AIE obtained using the Org1 particle formation mechanism is also enhanced in the 1750 atmosphere (−0.95 W $m^{-2}$ as compared to −0.77 W $m^{-2}$ in the present day; Table 4.3). The 1750 annual mean first AIE is more negative at most latitudes (dashed red line in Fig. 4.10), but between 30 and

**Table 4.3** Global annual mean changes to surface-level CCN (at 0.2 % supersaturation), CDNC at cloud height (approximately 900 hPa), and first AIE in a pre-industrial atmosphere, relative to an equivalent control simulation including no BVOC emission

| Experiment No. | Experiment name | ΔCCN ($cm^{-3}$) | ΔCDNC ($cm^{-3}$) | First AIE ($W\ m^{-2}$) |
|---|---|---|---|---|
| 22 | ACT_1750_mi | +23.8 (+23.2 %) | +12.8 (+12.6 %) | −0.19 |
| 23 | ACT_1750_mi_x0.5 | +17.8 (+17.3 %) | +10.7 (+10.6 %) | −0.18 |
| 24 | ACT_1750_mi_x2 | +30.5 (+29.7 %) | +14.7 (14.5 %) | −0.19 |
| 25 | ACT_1750_mi_x5 | +38.9 (+37.9 %) | +16.7 (+16.4 %) | −0.17 |
| 27 | Org1_1750_m | +52.5 (+56.7 %) | +34.9 (+37.9 %) | −0.95 |

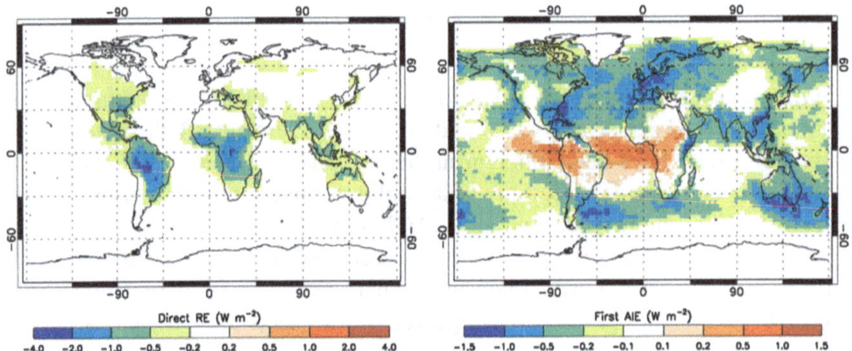


**Fig. 4.9** Annual mean DRE (*left*) and first AIE (*right*) from biogenic SOA when anthropogenic emissions from 1750 are included in GLOMAP (*ACT_1750_mi*)

50°S, the radiative effect is stronger in the present-day simulation (full red line Fig. 4.10) due to higher $SO_2$ emissions in southern Africa and Australia, required to generate the $H_2SO_4$ needed to form new particles using the Org1 nucleation mechanism.

Due to the lower background CCN concentration, increasing the SOA yield when 1750 anthropogenic emissions are used results in a greater fractional enhancement to CCN than when present-day emissions are used (Fig. 4.11). The enhanced CCN sensitivity to changes in SOA yield translates into a greater CDNC and local AIE sensitivity which has implications for the first aerosol indirect radiative forcing (RF) due to anthropogenic aerosol emissions. To test this, an anthropogenic first indirect RF since 1750 was calculated by setting $CDNC_1$ in Eq. 4.2 to the present day, and $CDNC_2$ to the value obtained from a simulation using anthropogenic emissions from 1750.

If a low source of biogenic SOA (18.5 Tg(SOA) $a^{-1}$) is assumed, the first aerosol indirect RF from anthropogenic emission changes (1750 to present) is $-1.16\ W\ m^{-2}$, but decreases in absolute value to $-1.10\ W\ m^{-2}$ when a large

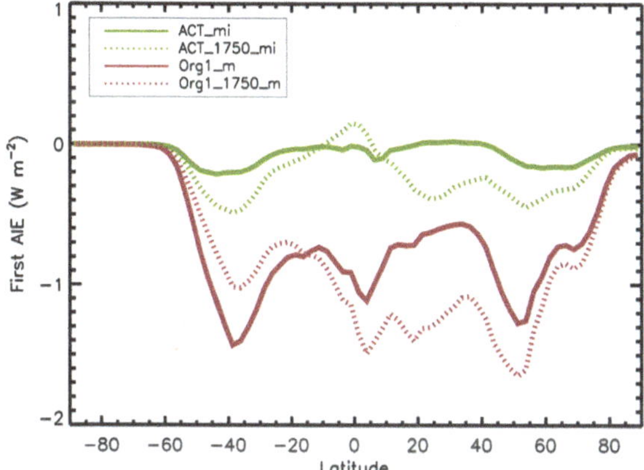


**Fig. 4.10** Annual zonal mean AIE from biogenic SOA in the present day (*solid lines*) and pre-industrial (*dashed lines*) when the ACT (*green*) and Org1 (*red*) nucleation mechanisms are used

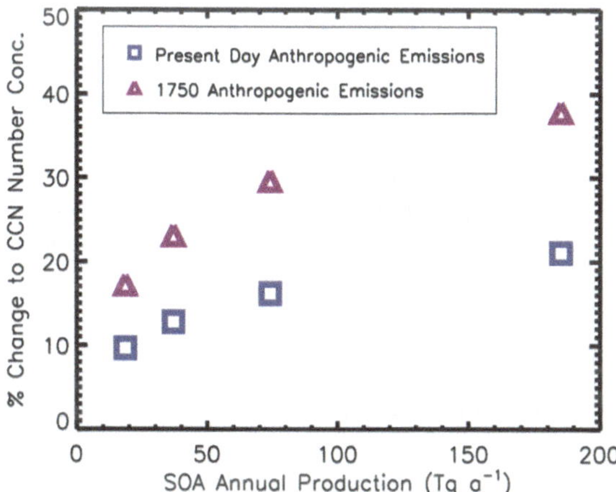


**Fig. 4.11** Percentage change to global annual mean CCN concentration, when biogenic SOA is included, simulated using the ACT nucleation mechanism at four different SOA production yields, with present-day (*blue squares*) and pre-industrial (*purple triangles*) anthropogenic emissions

source (185.1 Tg(SOA) $a^{-1}$) is assumed (Fig. 4.12). If the Org1 mechanism is used in both present-day and 1750 simulations (red circle in Fig. 4.12), the calculated aerosol indirect RF is $-1.04$ W $m^{-2}$; that is 0.12 W $m^{-2}$ smaller than that derived using the ACT mechanism.

This variation in the first indirect anthropogenic RF, due to uncertainties in the amount of SOA available and its behaviour in the atmosphere, highlights the need to understand the baseline pre-industrial atmosphere and the magnitude of pre-existing natural radiative effects in order to constrain the radiative forcings due to

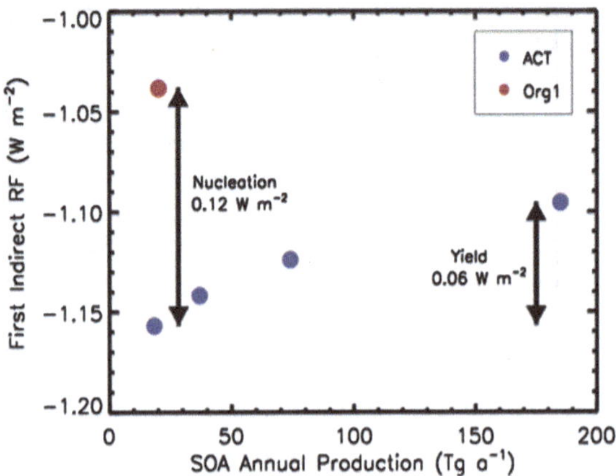


**Fig. 4.12** Anthropogenic first aerosol indirect RF from 1750 to present day, simulated using the ACT nucleation mechanism at four different yields for SOA production (*purple circles*), and using the Org1 nucleation mechanism with standard SOA production yield (*red circle*). *Arrows* highlight the RF sensitivity to assumptions about SOA yield and nucleation mechanism

human activities. Schmit et al. [26] demonstrated that uncertainty in volcanic $SO_2$ emissions plays a similar role in driving uncertainty in the anthropogenic first aerosol indirect RF.

In contrast, the DRE of biogenic SOA is less sensitive to the presence of anthropogenic emissions. When the ACT mechanism is used (Fig. 4.9, left), the same DRE is simulated with present-day and 1750 anthropogenic emissions ($-0.18$ W $m^{-2}$).

## 4.5 Summary and Conclusions

The inclusion of biogenic SOA leads to an increase (between 1.9 and 26.6 %) in the global annual mean CDNC, calculated offline. The spatial changes to CDNC from the inclusion of biogenic SOA broadly match changes to CCN concentrations; however, the magnitude of CDNC change becomes limited by competition for water vapour in highly polluted regions, such as those affected by biomass burning.

The inclusion of biogenic SOA results in a present-day global annual mean top-of-atmosphere DRE of between $-0.08$ and $-0.78$ W $m^{-2}$. The DRE is most sensitive to the yield of SOA production and strongest in the tropics where there are high BVOC emissions and high insolation. In Chap. 3, it was shown that altering (either increasing or decreasing) the yield of SOA production reduces the agreement with observations of $N_{80}$ at three forested locations and the subset of CCN observations used (Sect. 3.4); the best agreement was found for a global production of 37 Tg(SOA) $a^{-1}$, which gives a DRE from biogenic SOA of $-0.18$ W $m^{-2}$.

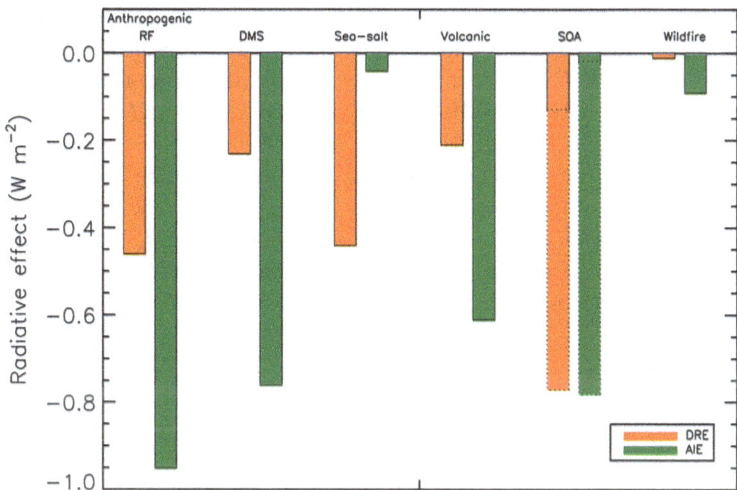


**Fig. 4.13** Anthropogenic indirect RF and REs from natural aerosol sources (*DMS* dimethyl sulphide, *SOA* biogenic secondary organic aerosol); values taken from Rap et al. [24], SOA bars are extended with the values from this chapter

The inclusion of biogenic SOA results in a present-day global annual mean top-of-atmosphere first AIE of between +0.01 and −0.77 W m$^{-2}$. The largest uncertainty in the AIE of biogenic SOA comes from the representation of new particle formation, specifically whether BVOC oxidation products contribute to the nucleation rate in the boundary layer. Including the oxidation products of monoterpenes in the particle formation rate equation gives up to an 11 times greater AIE, from biogenic SOA, than when $H_2SO_4$ alone controls new particle formation. The best agreement between simulated and measured seasonal cycles in $N_{80}$ is obtained when monoterpene oxidation products affect the new particle formation rate, suggesting that the magnitude of the global annual mean AIE from biogenic SOA could lie towards the most negative values in the estimated range presented here.

At high northern latitudes, monoterpene emissions from boreal forests result in summertime regional AIE of up to −5 W m$^{-2}$ over land, and −8 W m$^{-2}$ over ocean, when organic compounds influence the nucleation rate. This has implications for the overall climatic impact of high latitude forests and will be investigated in Chap. 6.

Figure 4.13 shows the REs calculated for a variety of natural aerosol sources by Rap et al. [24], indicating that previously, the strongest natural DRE came from sea-salt and the strongest natural AIE from DMS and volcanic eruptions. The size of the DRE and AIE bars for SOA in Fig. 4.13 are extended to represent the values with the largest absolute magnitude (−0.78 W m$^{-2}$ for the DRE and −0.77 W m$^{-2}$ for the AIE) calculated in this chapter. The extended bars indicate that at the outside of the uncertainty range examined here and in Chap. 3, biogenic SOA could exert the largest REs of the natural aerosol sources examined by Rap et al. [24].

When present-day anthropogenic aerosol emissions are replaced with those from 1750, the lower background aerosol concentrations result in a greater proportion of additional CCN becoming activated and therefore a more substantial AIE from biogenic SOA (−0.19 and −0.95 W $m^{-2}$ for *ACT_mi* and *Org1_m* respectively). As such, the AIE from biogenic SOA in 1750 is more sensitive to changes in the amount of SOA generated, and the nucleation mechanism used; adding uncertainty of 0.06 and 0.12 W $m^{-2}$ respectively to the magnitude of the first aerosol indirect RF from anthropogenic emissions since 1750 (Fig. 4.12). This highlights the need to understand the natural "background" state of the atmosphere in order to accurately quantify the impact of human activities.

## References


1. Abdul-Razzak H et al (1998) A parameterization of aerosol activation: 1. Single aerosol type. J Geophys Res Atmos 103(D6):6123–6131
2. Barahona D et al (2010) Comprehensively accounting for the effect of giant CCN in cloud activation parameterizations. Atmos Chem Phys 10(5):2467–2473
3. Bellouin N et al (2013) Impact of the modal aerosol scheme GLOMAP-mode on aerosol forcing in the Hadley Centre Global Environmental Model. Atmos Chem Phys 13(6):3027–3044
4. Bellouin N et al (2011) Aerosol forcing in the Climate Model Intercomparison Project (CMIP5) simulations by HadGEM2-ES and the role of ammonium nitrate. J Geophys Res 116(D20):D20206
5. Boucher O, Lohmann U (1995) The sulfate-CCN-cloud albedo effect. Tellus B 47(3):281–300
6. Bower KN et al (1994) A parameterization of warm clouds for use in atmospheric general circulation models. J Atmos Sci 51(19):2722–2732
7. Edwards JM, Slingo A (1996) Studies with a flexible new radiation code. I: choosing a configuration for a large-scale model. Q J R Meteorol Soc 122(531):689–719
8. Forster P et al (2007) Changes in atmospheric constituents and in radiative forcing. In: Climate change 2007: the physical science basis. Solomon S, Qin D, Manning M et al (eds) Contribution of Working Group I to the fourth assessment report of the Intergovernmental Panel on Climate Change. Cambridge University Press, Cambridge, UK and New York, USA
9. Fountoukis C, Nenes A (2005) Continued development of a cloud droplet formation parameterization for global climate models. J Geophys Res 110(D11):D11212
10. Goto D et al (2008) Importance of global aerosol modeling including secondary organic aerosol formed from monoterpene. J Geophys Res 113(D7):D07205
11. Guibert S et al (2003) Aerosol activation in marine stratocumulus clouds: 1. Measurement validation for a closure study. J Geophys Res 108(D15):8628
12. Jimenez JL et al (2009) Evolution of organic aerosols in the atmosphere. Science 326(5959):1525–1529
13. Jones A et al (1994) A climate model study of indirect radiative forcing by anthropogenic sulphate aerosols. Nature 370:450–453
14. Kurten T et al (2003) Estimation of different forest-related contributions to the radiative balance using observations in southern Finland. Boreal Environ Res 8:275–285
15. Lathière J et al (2005) Past and future changes in biogenic volatile organic compound emissions simulated with a global dynamic vegetation model. Geophys Res Lett 32:L20818
16. Lohmann U, Feichter J (2005) Global indirect aerosol effects: a review. Atmos Chem Phys 5(3):715–737

17. Mahowald N et al (2011) Aerosol impacts on climate and biogeochemistry. Annu Rev Environ Resour 36(1):45–74
18. Merikanto J et al (2010) Effects of boundary layer particle formation on cloud droplet number and changes in cloud albedo from 1850 to 2000. Atmos Chem Phys 10(2):695–705
19. Nenes A, Seinfeld JH (2003) Parameterization of cloud droplet formation in global climate models. J Geophys Res 108(D14):4415
20. O'Donnell D et al (2011) Estimating the direct and indirect effects of secondary organic aerosols using ECHAM5-HAM. Atmos Chem Phys 11(16):8635–8659
21. Peng Y et al (2005) Importance of vertical velocity variations in the cloud droplet nucleation process of marine stratus clouds. J Geophys Res 110(D21):D21213
22. Pringle KJ et al (2009) The relationship between aerosol and cloud drop number concentrations in a global aerosol microphysics model. Atmos Chem Phys 9(12):4131–4144
23. Ramaswamy V et al (2001) Radiative forcing. In: Climate change 2001: the physical science basis. Houghton JT, Ding Y, Griggs DJ et al (eds) Contribution of Working Group I to the third assessment report of the Intergovernmental Panel on Climate Change. Cambridge University Press, Cambridge, UK and New York, USA
24. Rap A et al (2013) Natural aerosol direct and indirect radiative effects. Geophys Res Lett 40:3297–3301
25. Rossow WB, Schiffer RA (1999) Advances in understanding clouds from ISCCP. B. Am. Meteorol. Soc. 80(11):2261–2287
26. Schmidt A et al (2012) Importance of tropospheric volcanic aerosol for indirect radiative forcing of climate. Atmos Chem Phys 12(16):7321–7339
27. Spracklen DV et al (2008) Boreal forests, aerosols and the impacts on clouds and climate. Philos Trans R Soc A 366:4613–4626
28. Spracklen DV et al (2011) Global cloud condensation nuclei influenced by carbonaceous combustion aerosol. Atmos Chem Phys 11(17):9067–9087
29. Twomey S (1977) Influence of pollution on shortwave albedo clouds. J Atmos Sci 34(7):1149–1152
30. Zhang Q et al (2007) Ubiquity and dominance of oxygenated species in organic aerosols in anthropogenically-influenced Northern Hemisphere midlatitudes. Geophys Res Lett 34(13):L13801

# Chapter 5
# The Impact of Volatility Treatment on the Radiative Effect of Biogenic SOA

## 5.1 Introduction

The multi-step oxidation of BVOCs yields products with lower volatility, which allows their partitioning to the particle phase and the formation of SOA. The manner in which this SOA adds to the existing aerosol distribution will influence its impact on the number, size and composition of particles in the atmosphere; in particular, the number of particles that are able to act as CCN.

As described in Chap. 1 and demonstrated in Chap. 3, the presence of SOA may increase CCN number concentrations by aiding the growth of smaller particles to CCN active sizes [22] and by making hydrophobic particles more hydrophilic [15]. Conversely, the presence of SOA may act to decrease CCN number concentrations by growing existing CCN sized particles to even larger sizes and enhancing the coagulational scavenging of ultrafine particles, and the condensational scavenging of potential nucleating gases; thereby suppressing new particle formation and growth as a route to CCN. The presence of these larger particles may also suppress the maximum in cloud supersaturation, allowing fewer potential CCN to become cloud droplets. Because the availability of CCN controls CDNC, and subsequently cloud albedo, the manner in which secondary organics are distributed has implications for the first AIE of biogenic SOA.

In this chapter, the radiative implications of two common approaches to modelling the behaviour of SOA will be quantified.

### *5.1.1 The Volatility Treatment of SOA*

The volatility, or vapour pressure, of a molecule is governed by its polarity and size. In the case of BVOCs, the addition of polar functional groups (i.e., through oxidation) will decrease their vapour pressure. The transfer of semi-volatile organic compounds between the gas and condensed phases can be treated using partitioning



C.E. Scott, *The Biogeochemical Impacts of Forests and the Implications for Climate Change Mitigation*, Springer Theses, DOI 10.1007/978-3-319-07851-9_5

theory that assumes instant equilibration between organic vapours and the organic mass in the aerosol phase [13, 14]. When simulating the aerosol size distribution, a consequence of this instant-equilibrium approach is that the net condensation of new organic mass scales with the existing organic mass size distribution of the particles [21]. Because aerosol mass scales with volume, this means that particles requiring condensation to grow to climatically relevant sizes receive only a trivial fraction of the new SOA and subsequently do not grow. However, if the volatility of organic oxidation products is further reduced (i.e., through gas or particle-phase chemistry [4, 9]), they may condense kinetically according to the Fuchs-corrected surface area of existing particles and a larger proportion of the condensable mass will be added to the nucleation mode [21, 28, 29].

Neither approach fully describes the behaviour of SOA; the kinetic approach neglects the re-evaporation of semi-volatile organics whilst the thermodynamic approach is unable to account for the observed growth in particles smaller than 100 nm in diameter [17, 18, 21, 28].

Global aerosol microphysics models use either the thermodynamic (partitioning proportional to organic mass, e.g., [2, 8, 12, 19, 26]) or the kinetic (condensation proportional to particle surface area e.g. [10, 25]) assumptions described above. Riipinen et al. [21] and D'Andrea et al. [3] both found that the simulated global annual mean concentration of CCN-sized particles increased by approximately 10 % when the kinetic (rather than thermodynamic) assumption was used, with regional increases of over 50 %. Yu [28] found that allowing successive stages of oxidation, and the generation of non-volatile products, to occur increased simulated CCN concentrations by 5–50 % at the surface, over a version of the same model in which the thermodynamic assumption was applied.

These global aerosol microphysics models have been used to quantify the cloud albedo, or first indirect effect (AIE), of biogenic SOA, estimating values that span from positive (e.g. +0.23 W $m^{-2}$; [12]) to negative (−0.02 W $m^{-2}$; [20], and e.g. −0.07 W $m^{-2}$; Chap. 4 of this thesis). One difference between these studies is the method by which they represent the condensation of SOA, with O'Donnell et al. [12] applying a thermodynamic approach whereas the other studies use the kinetic approach.

In this chapter, the implications of these two different approaches, used to distribute secondary organics amongst the existing aerosol population, for the sign and magnitude of the aerosol indirect effect from biogenic SOA will be examined. The direct radiative effect for a given aerosol component tends to scale linearly with the mass of particulate material (e.g. [20]) and may be less sensitive to the relative proportion of ultrafine and larger particle growth.

## 5.2 Experimental Setup

In this chapter, two different global aerosol microphysics models are used to test the hypothesis that the sign of the first AIE of biogenic SOA is controlled by the manner in which secondary organic material is distributed across the aerosol size distribution.

The first global aerosol microphysics model used in this analysis is GLOMAP-mode, described in Chap. 2 and used in Chap. 3. The second model is the global 3D chemical transport model GEOS-Chem (v8.02.02) (http://www.geos-chem.org) coupled with the TwO-Moment Aerosol Sectional (TOMAS) microphysics model [1, 16, 24]. GEOS-Chem-TOMAS (referred to as "TOMAS" throughout the rest of this chapter) operates at a horizontal resolution of $4^{\circ} \times 5^{\circ}$ with 30 $\sigma$-pressure levels from the surface to 0.01 hPa. In contrast to GLOMAP-mode, TOMAS uses 40 log-normally spaced bins to simulate particles with diameters between 1 nm and 10 μm. The TOMAS simulations in this chapter were performed by Jeffrey Pierce; all subsequent analysis was performed by the candidate.

The GLOMAP-mode experiments examined in this chapter are equivalent to the *ACT_m* experiment described in Chap. 3; i.e., include BHN and an empirically derived mechanism for the activation rate of sulphuric acid clusters in the boundary layer (Sect. 2.1.4.1). In TOMAS, the same activation mechanism is used in the boundary layer, but BHN is parameterised according to Vehkamäki et al. [27].

In TOMAS, secondary organic material is generated at a fixed molar yield of 10 % (as compared to 13 % in GLOMAP), from the oxidation of monoterpenes by $O_3$, OH and $NO_3$. In contrast to GLOMAP, which takes monoterpene emissions from the GEIA database, TOMAS uses monoterpene emissions generated using the Model of Emissions of Gases and Aerosols from Nature (MEGAN) v2.0 [6], generating 18.4 Tg(SOA) $a^{-1}$ (as compared to 20.4 Tg(SOA) $a^{-1}$ in GLOMAP).

In this chapter, SOA is distributed across the aerosol size distribution using two different approaches. In both models the standard approach (described in Sect. 2.1. 4.3) is to assume secondary organic mass ($M_{SOA}$) condenses as if it were non-volatile, to the Fuchs-corrected surface area (i.e., the kinetic approach), such that the rate of change of $M_{SOA}$ in each mode/bin $i$, may be described as in Eq. 5.1, where $S_{org}$ represents the gas-phase concentration of secondary organic material:

$$\frac{dM_{SOA_i}}{dt} = \frac{C_i N_i}{\sum_{i=1,5} C_i N} \times \frac{dS_{org}}{dt} \tag{5.1}$$

where $C_i$ represents the condensation coefficient for each mode $i$ (Eq. 2.8) and $N_i$ is the particle number concentration. Here, a sensitivity study is conducted in which the amount of secondary organic material entering the aerosol phase is partitioned between the size modes/bins according to Eq. 5.2, where $M_{OAi}$ is the pre-existing organic mass in mode $i$ (i.e., the thermodynamic assumption):

$$\frac{dM_{SOA_i}}{dt} = \frac{M_{OA_i}}{\sum_{i=1,5} M_{OA_i}} \times \frac{dS_{org}}{dt} \tag{5.2}$$

Because the aim is to quantify the impact of changes in the size of particles to which the SOA condenses, SOA is otherwise treated identically between the two different approaches and is not allowed to re-partition into the gas phase.

With each model one simulation without biogenic SOA (*NoSOA*) is completed, as well as two simulations with biogenic SOA (*KinSOA* and *ThermSOA*). The resulting changes to CDNC due to the inclusion of biogenic SOA (i.e., *KinSOA–NoSOA* or *ThermSOA–NoSOA*) are calculated using the parameterisation developed by Nenes and Seinfeld [11] and described in Chap. 4, with a uniform updraught velocity of 0.2 m $s^{-1}$. The first AIE of biogenic SOA is then calculated using the offline radiative transfer model of Edwards and Slingo [5], and the approach described in Chap. 4. The global distribution of cloud updraught velocity is uncertain; to determine the sensitivity of these results to updraught velocity, the first AIE is calculated for a range of globally uniform updraught values (0.1–0.5 m $s^{-1}$).

## 5.3 Results

Figure 5.1 compares the observed aerosol size distributions at a boreal forest location (Hyytiälä, Finland), with those simulated by GLOMAP and TOMAS using the kinetic and thermodynamic approaches. At Hyytiälä, the simulated size distribution will be sensitive to many processes including new particle formation, the amount of SOA, and the characteristics of primary particles; the intention here is to demonstrate that the aerosol size distribution is also sensitive to the treatment of SOA condensation.

The number of particles between 40 and 200 nm in diameter is underestimated by both models, in every simulation (Fig. 5.1). Results from Chap. 3 suggest that organically mediated new particle formation (not included here) may be required to accurately simulate the number of particles between 20 and 100 nm (Fig. 3.8). In the absence of SOA, a large nucleation mode is simulated by both models (dashed blue lines in Fig. 5.1). In GLOMAP, when the kinetic approach is applied (green lines in Fig. 5.1), some secondary organic material condenses onto particles in the nucleation mode (0.26 % of the total flux in GLOMAP), enabling their growth into the 40–200 nm size range. This is consistent with Riipinen et al. [21] who found that variation in the growth rate of particles between 7 and 20 nm was linked to the presence of organic oxidation products.

When the thermodynamic approach is applied (red line in Fig. 5.1), no secondary organic material is added to the nucleation mode, suppressing the growth of these particles. Rather, the secondary organic material is added to particles in the Aitken and accumulation modes with greater existing organic mass.

Table 5.1 summarises the flux of secondary organic material to particles of different sizes in GLOMAP. Using the thermodynamic approach, particles smaller than 100 nm diameter receive less than half the amount of secondary organic material per particle ($2.97 \times 10^{-12}$ ng(SOA) $\text{particle}^{-1}$ $s^{-1}$), as compared to the kinetic approach ($6.77 \times 10^{-12}$ ng(SOA) $\text{particle}^{-1}$ $s^{-1}$). This response is consistent with Yu [28] and with D'Andrea et al. [3] who observed a similar relative response across 20 ground-based measurement sites, when comparing a kinetic and thermodynamic approach.

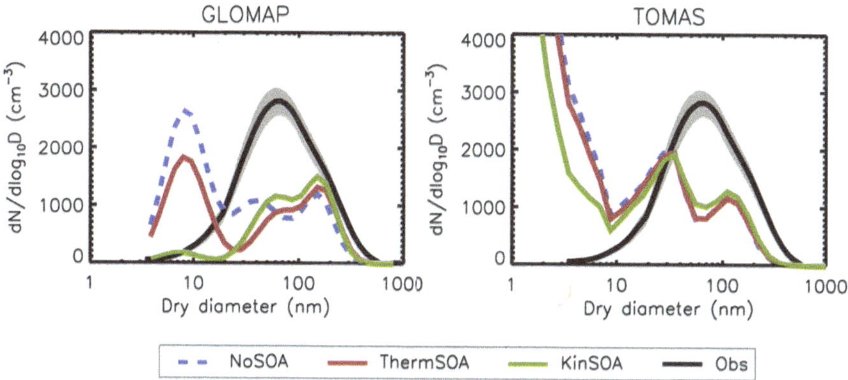


**Fig. 5.1** Simulated (GLOMAP (*left*) and TOMAS (*right*)) and measured (multi-annual; 1996–2006) mean size distribution at Hyytiälä during June–July–August. The *grey* shaded region represents the mean (over June–July–August) standard deviation of the annual mean size distributions

### *5.3.1 Changes to Cloud Droplet Number Concentration*

Figure 5.2 shows the simulated change in annual mean CDNC due to the presence of biogenic SOA in both models, using an updraught velocity of 0.2 m $s^{-1}$. The spatial pattern of CDNC change, upon inclusion of biogenic SOA, varies between the two models.

As in Chap. 4, under the kinetic approach, GLOMAP simulates the largest fractional increases to CDNC over boreal regions and southern hemisphere oceans (Fig. 5.2a). In TOMAS, the largest increases are simulated over tropical land regions (Fig. 5.2c). Over most regions, GLOMAP simulates higher background CDNC than TOMAS, with the annual global mean *NoSOA* CDNC approximately 60 % ($\sim$70 $cm^{-3}$) higher in GLOMAP. However, annual mean CDNC in TOMAS are up to 70 % ($\sim$70 $cm^{-3}$) higher over some continental boreal regions of Canada and Siberia, than GLOMAP CDNC; therefore, a small absolute CDNC change over the boreal region in GLOMAP results in a relatively larger fractional effect. Additionally, monoterpene emission rates in the GEIA inventory (used here in GLOMAP) are up to a factor of 10 higher over some boreal regions than those generated by MEGANv2.0 (used here in TOMAS) [23]; this is discussed in further detail in Chap. 6.

Over tropical regions, particle concentrations are dominated by biomass burning emissions during the dry season. Both models use primary carbonaceous emissions from the GFED inventory, but these are emitted with a number median diameter of 150 nm in GLOMAP (see Sect. 2.1.3), and 100 nm in TOMAS. This means that for the same mass of biomass burning emission, a greater number of smaller particles (that require SOA for growth to CCN sizes) are emitted in TOMAS over tropical regions, than in GLOMAP. Additionally, monoterpene emissions generated by MEGANv2.0 are up to 100 % higher over the tropics than those

**Table 5.1** Global annual mean fluxes of secondary organic material, in the model surface layer, to the aerosol size distribution in GLOMAP

| Distribution of secondary organic material | | Smaller than 100 nm | | Larger than 100 nm | |
|---|---|---|---|---|---|
| | | Nucleation ($D_g < 10$ nm) | Aitken ($10 < D_g < 100$ nm) | Accumulation ($100$ nm $< D_g < 1$ μm) | Coarse ($D_g > 1$ μm) |
| Kinetic | % of total flux | 0.26 | 20.94 | 78.26 | 0.55 |
| | Total flux | 4.28 Tg(SOA) $a^{-1}$<br>$6.77 \times 10^{-12}$ ng(SOA) particle$^{-1}$ $s^{-1}$ | | 15.92 Tg(SOA) $a^{-1}$<br>$1.50 \times 10^{-11}$ ng(SOA) particle$^{-1}$ $s^{-1}$ | |
| Thermodynamic | % of total flux | 0.00 | 21.12 | 78.85 | 0.03 |
| | Total flux | 3.79 Tg(SOA) $a^{-1}$<br>$2.97 \times 10^{-12}$ ng(SOA) particle$^{-1}$ $s^{-1}$ | | 14.17 Tg(SOA) $a^{-1}$<br>$1.74 \times 10^{-11}$ ng(SOA) particle$^{-1}$ $s^{-1}$ | |

$D_g$ is the geometric mean diameter for each mode

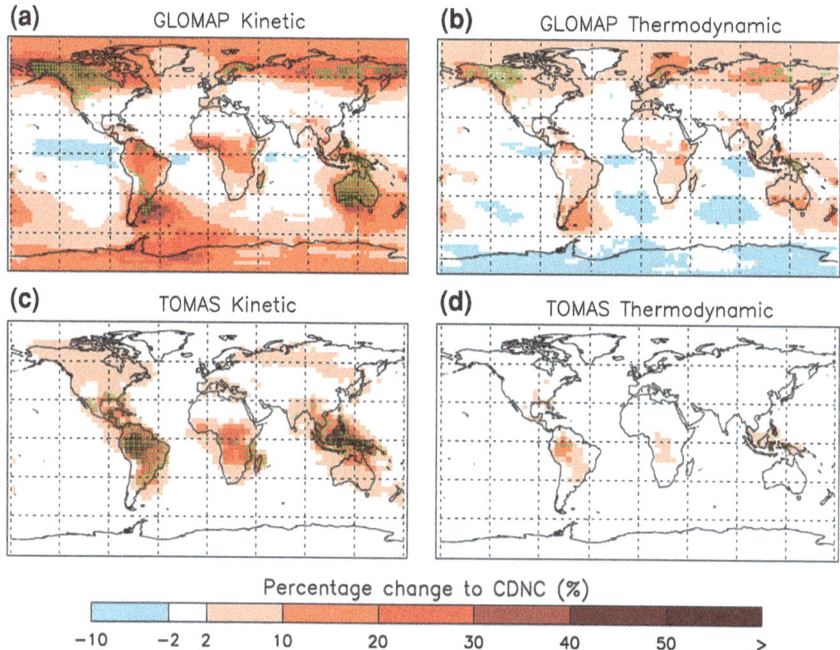


**Fig. 5.2** Annual mean percentage change to CDNC (using a uniform updraught velocity of 0.2 m $s^{-1}$) from biogenic SOA in GLOMAP (*upper*) and TOMAS (*lower*), in the model level which corresponds to low level cloud base (mean pressure of approximately 900 hPa). Secondary organic mass is distributed according to the kinetic approach in panels (**a**) and (**c**), and according to the thermodynamic approach in panels (**b**) and (**d**). *Green crosses* denote grid cells with more than a 30 % increase in particles with diameter greater than 80 nm ($N_{80}$) when biogenic SOA is included

from the GEIA inventory [7]; discussed further in Chap. 6. The additional supply of ultrafine particles, and greater amount of SOA produced in the tropics by TOMAS leads to a higher CDNC sensitivity to biogenic SOA in this region. Since the background CDNC over the tropics is higher in GLOMAP, which reduces the maximum supersaturation, this will allow fewer additional CCN sized particles to form cloud droplets.

As was also seen in Chap. 4, the inclusion of biogenic SOA in GLOMAP results in reductions in CDNC over some oceanic regions downwind of biomass burning emission regions (Fig. 5.2a). This reduction in CDNC is due to the condensation of SOA onto hydrophobic biomass burning aerosols (i.e. physical ageing), enhancing their loss rate by wet deposition. In the *NoSOA* simulations, this process takes longer and these particles may survive to act as CCN over the oceanic regions. The process of explicitly ageing hydrophobic biomass burning emissions by SOA is not represented in TOMAS (aerosol ageing occurs on a fixed 1.5 day timescale), so the reduction in CDNC over the oceans is not seen in Fig. 5.2c.

When the thermodynamic approach is applied, the secondary organic material is preferentially distributed towards the aerosol modes/bins with the greatest existing mass of organic material, thereby increasing the size of larger particles. Consequently, the growth of ultrafine particles is lower than with the kinetic approach, and far fewer model grid cells (in both models, Fig. 5.2) experience more than a 30 % increase in the number of particles with dry diameter greater than 80 nm ($N_{80}$). As a result, regional annual mean increases in CDNC are limited to approximately 30 % in both models (Fig. 5.2b, d). In GLOMAP, regions of negative CDNC change are larger than with the kinetic approach because the increased scavenging of hydrophobic particles is not balanced by the growth of ultrafine particles as it was in the kinetic approach.

Table 5.2 reports the change in global annual mean CDNC when biogenic SOA is included in both models. Across all five updraught velocities (0.1–0.5 m $s^{-1}$), and in both models, the kinetic approach leads to a larger global annual mean increase in CDNC than the thermodynamic approach. Increasing the updraught velocity increases the absolute and fractional change in CDNC at cloud base due to biogenic SOA. However, the relative kinetic to thermodynamic response remains consistent across the five updraught velocities (Table 5.2).

### *5.3.2 First Aerosol Indirect Effect*

Figure 5.3 shows the spatial variation in annual mean first AIE (when CDNC have been calculated using an updraught velocity of 0.2 m $s^{-1}$), with the kinetic and thermodynamic approach, for both models. The kinetic approach gives biogenic SOA a negative global annual mean first AIE in GLOMAP (−0.07 W $m^{-2}$; updraught velocity of 0.2 m $s^{-1}$), with regions of annual mean negative forcing peaking at high northern latitudes (~60°N), the tropics, and southern hemisphere (30–50°S). The small CDNC decreases seen over the tropical oceans in Fig. 5.2a result in a positive radiative effect (Fig. 5.3a), reducing the magnitude of the first AIE at 0° latitude (Fig. 5.4, *upper*). In TOMAS, the kinetic approach gives a global annual mean first AIE of −0.03 W $m^{-2}$ (for all updraught velocities; Table 5.2), peaking at tropical latitudes where the greatest increase in CDNC is simulated (Fig. 5.2c).

Figure 5.4 shows the zonal mean first AIE, across the five updraught velocities, for each treatment in both models. In GLOMAP, taking the thermodynamic approach gives a global annual mean first AIE of +0.02 W $m^{-2}$ (with an updraught velocity of 0.2 m $s^{-1}$), with the zonal mean peaking at around 0° latitude (Fig. 5.4, *upper*). This occurs due to decreases in CDNC at the height of low-level (Fig. 5.2b) and mid-level (Fig. 5.5, *upper*) clouds in the tropics. The higher altitude (~600 hPa) decrease occurs as a result of the enhanced condensation sink for vapours (e.g. $H_2SO_4$) at the surface (due to larger particle size), and subsequent

**Table 5.2** Global annual mean change to cloud droplet number concentration (CDNC), calculated using five globally uniform updraught velocities

| Model | Updraught velocity ($m\ s^{-1}$) | Distribution of secondary organic material | | | | Ratio of CDNC change: $\Delta CDNC_{kin}/\Delta CDNC_{therm}$ |
|---|---|---|---|---|---|---|
| | | Kinetic | | Thermodynamic | | |
| | | ΔCDNC ($cm^{-3}$) | First AIE ($W\ m^{-2}$) | ΔCDNC ($cm^{-3}$) | First AIE ($W\ m^{-2}$) | |
| GLOMAP | 0.1 | +2.7 (+2.0 %) | −0.05 | +0.7 (+0.6 %) | +0.03 | 3.6 |
| | **0.2** | **+6.3 (+3.4 %)** | **−0.07** | **+2.1 (+1.1 %)** | **+0.02** | **3.0** |
| | 0.3 | +9.1 (+4.3 %) | −0.08 | +3.1 (+1.5 %) | +0.03 | 2.9 |
| | 0.4 | +11.4 (+4.9 %) | −0.08 | +3.9 (+1.7 %) | +0.03 | 2.9 |
| | 0.5 | +12.9 (+5.3 %) | −0.08 | +4.4 (+1.8 %) | +0.03 | 2.9 |
| | **Mean** | **+8.4 (+4.0 %)** | **−0.07** | **+2.9 (+1.3 %)** | **+0.03** | **3.0** |
| TOMAS | 0.1 | +2.7 (+3.2 %) | −0.03 | +0.3 (+0.3 %) | 0.00 | 9.6 |
| | **0.2** | **+3.9 (+3.4 %)** | **−0.03** | **+0.3 (+0.3 %)** | **0.00** | **11.4** |
| | 0.3 | +5.0 (+3.7 %) | −0.03 | +0.5 (+0.3 %) | 0.00 | 10.5 |
| | 0.4 | +5.9 (+3.9 %) | −0.03 | +0.5 (+0.4 %) | 0.00 | 10.6 |
| | 0.5 | +6.6 (+4.0 %) | −0.03 | +0.6 (+0.4 %) | 0.00 | 10.8 |
| | **Mean** | **+4.8 (+3.6 %)** | **−0.03** | **+0.5 (+0.3 %)** | **0.00** | **10.6** |

In the model level which corresponds to low-level cloud base (mean pressure of approximately 900 hPa), and first aerosol indirect effect (AIE), reported to 2 decimal places, resulting from the inclusion of biogenic SOA in GLOMAP and TOMAS simulations using the kinetic and thermodynamic approaches

A mean ΔCDNC and first AIE for each model is calculated, assuming that each updraught velocity is equally likely to occur

suppression of binary nucleation in the free troposphere. In TOMAS, a negligible global annual mean first AIE (0.00 W $m^{-2}$ to 2 decimal places) is simulated with the thermodynamic approach due to the small change in CDNC (Table 5.2 and Fig. 5.2).

Due to the greater change in CDNC with increasing updraught velocity, the GLOMAP global annual mean first AIE becomes more negative (Table 5.2), reaching −0.08 W $m^{-2}$ with an updraught velocity between 0.3 and 0.5 m $s^{-1}$. When the thermodynamic approach is taken in GLOMAP, the global annual mean first AIE varies by less than 0.005 W $m^{-2}$ when the updraught velocity is modified

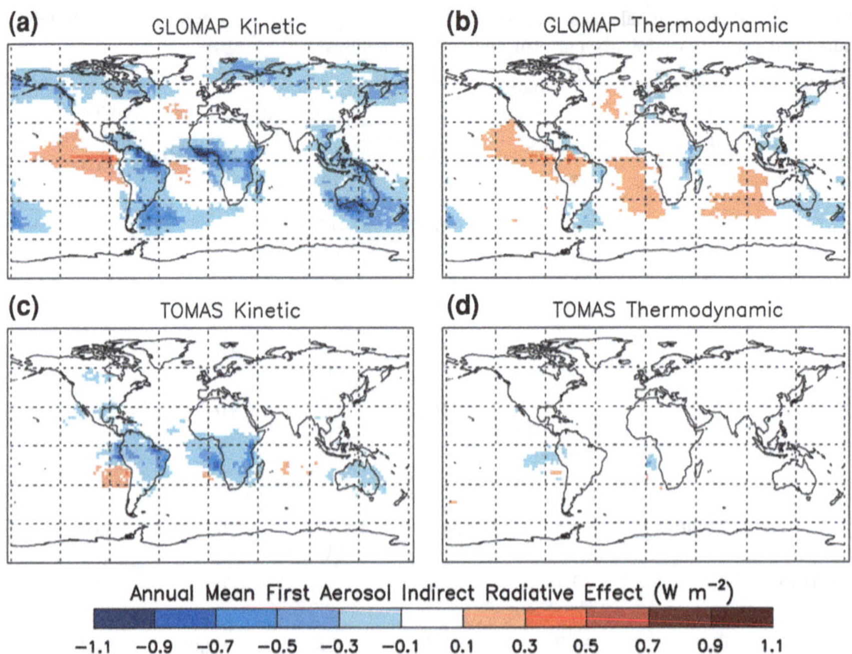


**Fig. 5.3** Annual mean first AIE (W m$^{-2}$) from biogenic SOA in GLOMAP (*upper*) and TOMAS (*lower*) when secondary organic mass distributed kinetically (**a** and **c**) and thermodynamically (**b** and **d**)

(Table 5.2), due to the enhancement of regions of both positive and negative AIE with increasing updraught velocity (Fig. 5.4).

When the kinetic approach is applied in TOMAS, and the updraught velocity is increased, the calculated first AIE remains consistent at −0.03 W m$^{-2}$ (Table 5.2), with much of the AIE simulated at tropical latitudes (Figs. 5.3c and 5.4, *lower*). This occurs because whilst the magnitude of the zonal mean CDNC increase at cloud base (over 20°S–20°N) increases, CDNC at the altitude of mid-level clouds decreases further (Fig. 5.5).

The global mean first AIE shows a strong seasonal cycle in GLOMAP when the kinetic approach is applied, peaking in August (Fig. 5.6) due to a substantial negative first AIE during NH summertime. In TOMAS, a negligible seasonal cycle is simulated when the kinetic approach is applied due to the dominance of CDNC change in the tropics, where monoterpene emissions are high throughout the year, and the less substantial change to CDNC at high northern latitudes as compared to GLOMAP. When the thermodynamic approach is applied, the first AIE calculated from both models shows a negligible seasonal cycle, again due to the dominance of tropical forcings (Fig. 5.3) and because regions of small positive and small negative first AIE balance throughout the year.

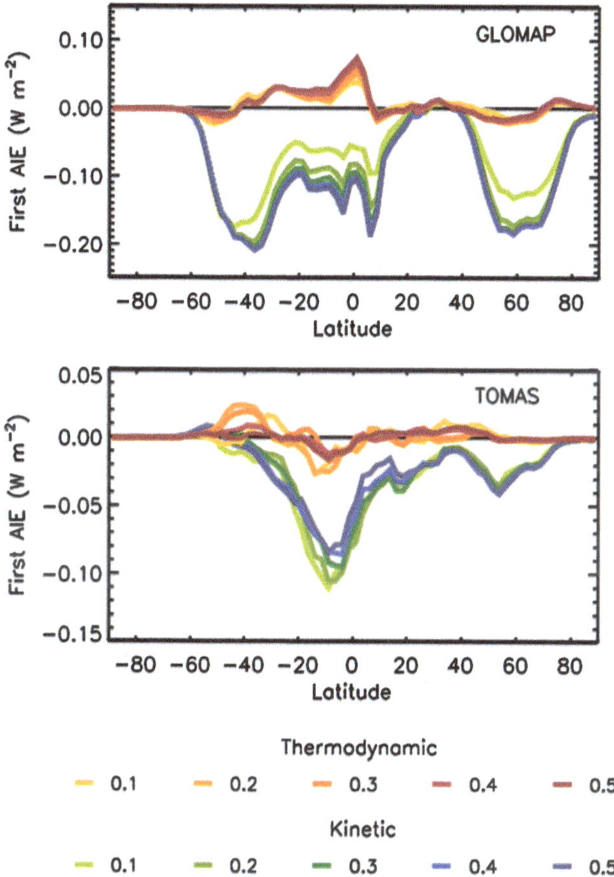


**Fig. 5.4** Annual zonal mean first AIE due to biogenic SOA in GLOMAP (*upper*) and TOMAS (*lower*), based on CDNC changes calculated at five different globally uniform updraught velocities (0.1–0.5 m $s^{-1}$); note different vertical axis scales

## 5.4 Summary and Conclusions

In this chapter, two different detailed microphysics models are used to examine the implication of volatility treatment on the radiative impact of biogenic SOA. Despite differences in the spatial distribution of CDNC changes between the two models, the magnitude of the CDNC response is strongly controlled by the size of particles onto which secondary organic material is distributed.

The kinetic approach, which enables organic oxidation products to condense upon the smallest particles, facilitating their growth to larger sizes, increased annual mean CDNC and gave a negative first AIE when biogenic SOA was included. Applying the thermodynamic approach suppresses the growth of the smallest particles; globally this resulted in a smaller increase to simulated CDNC

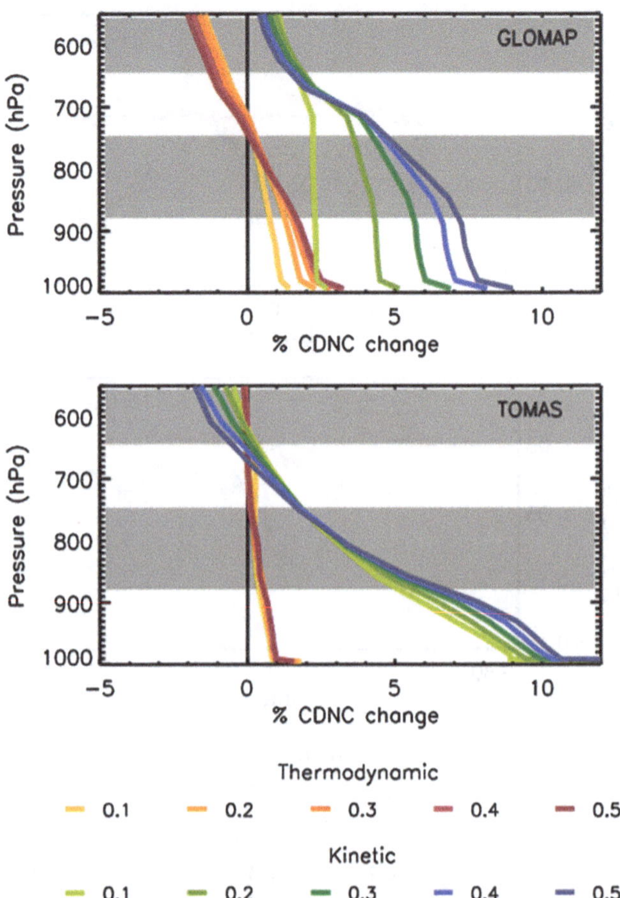


**Fig. 5.5** Annual zonal mean percentage CDNC change, over tropical latitudes (20°N–20°S), due to biogenic SOA in GLOMAP (*upper*) and TOMAS (*lower*), at five different globally uniform updraught velocities (between 0.1 and 0.5 m $s^{-1}$); *grey* shaded areas indicate pressure levels in which cloud fraction is greater than 10 % (in the ISCCP dataset for the year 2000) at these latitudes

when biogenic SOA was included, and gave a negligible or small positive first AIE. Despite structural differences in the two models used here, a consistent response has been found; indicating that these findings have general implications for global aerosol models.

In this chapter it has been shown that the approach used to model SOA controls the sign of the calculated first AIE. Accurately simulating the condensation of SOA onto ultrafine particles is important when evaluating processes that depend strongly on changes to ultrafine particle number, such as the AIE.

The current GLOMAP-approach partitions SOA entirely kinetically, and may therefore underestimate the amount of SOA condensing onto larger particles.

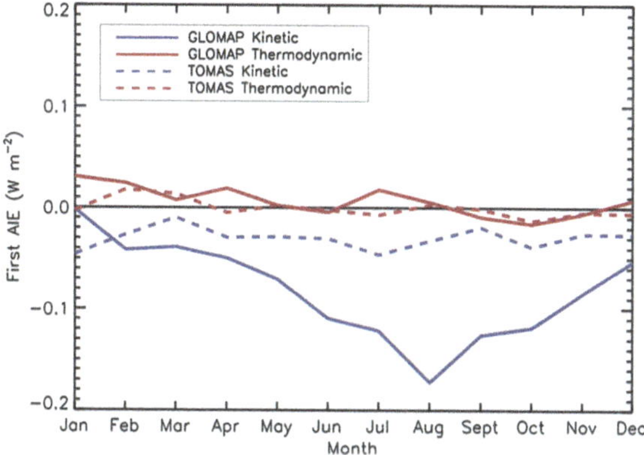


**Fig. 5.6** Seasonal cycle in global mean first AIE (W m$^{-2}$) simulated using GLOMAP (*solid lines*) and TOMAS (*dashed lines*); partitioning organics according to the kinetic model (*purple lines*) and thermodynamic model (*red lines*)

Ultimately, a combination of the thermodynamic and kinetic approaches will be required in order to accurately represent the range of differing volatility compounds present in the atmosphere. As gas- and particle-phase atmospheric chemistry tends to reduce the volatility of organic compounds, accounting for the presence of non-volatiles, as well as semi-volatiles, is important. Improving on existing first attempts (e.g., [21, 28]) will require a more detailed understanding of the pathways by which organic compounds of differing volatilities are generated, and their relative contributions to the growth of particles of different sizes.

## References

1. Adams PJ, Seinfeld JH (2002) Predicting global aerosol size distributions in general circulation models. J Geophys Res Atmos, 107(D19):AAC 4-1–AAC 4-23
2. Chung SH, Seinfeld JH (2002) Global distribution and climate forcing of carbonaceous aerosols. J Geophys Res 107(D19):4407
3. D'Andrea SD et al (2013) Understanding and constraining global secondary organic aerosol amount and size-resolved condensational behavior. Atmos Chem Phys 13:11519–11534
4. Donahue NM et al (2011) Theoretical constraints on pure vapor-pressure driven condensation of organics to ultrafine particles. Geophys Res Lett 38(16):L16801
5. Edwards JM, Slingo A (1996) Studies with a flexible new radiation code. I: choosing a configuration for a large-scale model. Q J Roy Meteorol Soc 122(531):689–719
6. Guenther A et al (2006) Estimates of global terrestrial isoprene emissions using MEGAN (Model of Emissions of Gases and Aerosols from Nature). Atmos Chem Phys 6(11):3181–3210
7. Guenther AB et al (2012) The Model of Emissions of Gases and Aerosols from Nature version 2.1 (MEGAN2.1): an extended and updated framework for modeling biogenic emissions. Geosci Model Dev 5(6):1471–1492

8. Heald CL et al (2008) Predicted change in global secondary organic aerosol concentrations in response to future climate, emissions, and land use change. J Geophys Res 113(D5):D05211
9. Jimenez JL et al (2009) Evolution of organic aerosols in the atmosphere. Science 326(5959):1525–1529
10. Makkonen R et al (2009) Sensitivity of aerosol concentrations and cloud properties to nucleation and secondary organic distribution in ECHAM5-HAM global circulation model. Atmos Chem Phys 9(5):1747–1766
11. Nenes A, Seinfeld JH (2003) Parameterization of cloud droplet formation in global climate models. J Geophys Res 108(D14):4415
12. O'Donnell D et al (2011) Estimating the direct and indirect effects of secondary organic aerosols using ECHAM5-HAM. Atmos Chem Phys 11(16):8635–8659
13. Odum JR et al (1996) Gas/particle partitioning and secondary organic aerosol yields. Environ Sci Technol 30(8):2580–2585
14. Pankow JF (1994) An absorption model of the gas/aerosol partitioning involved in the formation of secondary organic aerosol. Atmos Environ 28(2):189–193
15. Petters MD et al (2006) Chemical aging and the hydrophobic-to-hydrophilic conversion of carbonaceous aerosol. Geophys Res Lett 33(24):L24806
16. Pierce JR et al (2013) Weak global sensitivity of cloud condensation nuclei and the aerosol indirect effect to Criegee + SO2 chemistry. Atmos Chem Phys 13(6):3163–3176
17. Pierce JR et al (2012) Nucleation and condensational growth to CCN sizes during a sustained pristine biogenic SOA event in a forested mountain valley. Atmos Chem Phys 12(7):3147–3163
18. Pierce JR et al (2011) Quantification of the volatility of secondary organic compounds in ultrafine particles during nucleation events. Atmos Chem Phys 11(17):9019–9036
19. Pye HOT, Seinfeld JH (2010) A global perspective on aerosol from low-volatility organic compounds. Atmos Chem Phys 10(9):4377–4401
20. Rap A et al (2013) Natural aerosol direct and indirect radiative effects. Geophys Res Lett 40:3297–3301
21. Riipinen I et al (2011) Organic condensation: a vital link connecting aerosol formation to cloud condensation nuclei (CCN) concentrations. Atmos Chem Phys 11(8):3865–3878
22. Riipinen I et al (2012) The contribution of organics to atmospheric nanoparticle growth. Nat Geosci 5(7):453–458
23. Sakulyanontvittaya T et al (2008) Monoterpene and sesquiterpene emission estimates for the United States. Environ Sci Technol 42(5):1623–1629
24. Snow-Kropla EJ et al (2011) Cosmic rays, aerosol formation and cloud-condensation nuclei: sensitivities to model uncertainties. Atmos Chem Phys 11(8):4001–4013
25. Spracklen DV et al (2006) The contribution of boundary layer nucleation events to total particle concentrations on regional and global scales. Atmos Chem Phys 6(12):5631–5648
26. Tsigaridis K et al (2005) Naturally driven variability in the global secondary organic aerosol over a decade. Atmos Chem Phys 5(7):1891–1904
27. Vehkamäki H et al (2002) An improved parameterization for sulfuric acid–water nucleation rates for tropospheric and stratospheric conditions. J Geophys Res Atmos 107(D22):4622
28. Yu F (2011) A secondary organic aerosol formation model considering successive oxidation aging and kinetic condensation of organic compounds: global scale implications. Atmos Chem Phys 11(3):1083–1099
29. Zhang X et al (2012) Diffusion-limited versus quasi-equilibrium aerosol growth. Aerosol Sci Technol 46(8):874–885

# Chapter 6
# The Radiative Effects of Deforestation

## 6.1 Introduction

In Chap. 4, it was shown that the presence of biogenic SOA has a considerable radiative influence, when compared to other natural components of the atmosphere. In particular, the first AIE due to biogenic SOA may be substantial if biogenic oxidation products contribute to the initial stages of new particle formation.

In this chapter, a land-surface model is used to explore the impact of several idealised deforestation scenarios on BVOC emission levels. The subsequent change in aerosol properties is then calculated using GLOMAP-mode, and the first AIE is evaluated using the Edwards-Slingo (ES) radiative transfer model. To put the magnitude of this indirect radiative effect into context, the equivalent radiative effects of changes to surface albedo and atmospheric carbon dioxide concentration associated with deforestation are also evaluated (Fig. 6.1).

## 6.2 Experimental Setup

### *6.2.1 The Community Land Model and MEGAN*

The land component of the Community Atmosphere Model (CAM) and the Community Climate System Model (CCSM), known as the Community Land Model (CLMv4.0) [8, 13], was used to generate speciated BVOC emission fields, according to the Model of Emissions of Gases and Aerosols from Nature (MEGANv2.1) [5]. The CLM operates at a horizontal resolution of 2.5° (lon) $\times$ 1.9° (lat). The CLM simulations described in this chapter were performed by Stephen Arnold; the land cover files were modified by the candidate and all subsequent analysis was performed by the candidate.



C.E. Scott, *The Biogeochemical Impacts of Forests and the Implications for Climate Change Mitigation*, Springer Theses, DOI 10.1007/978-3-319-07851-9_6

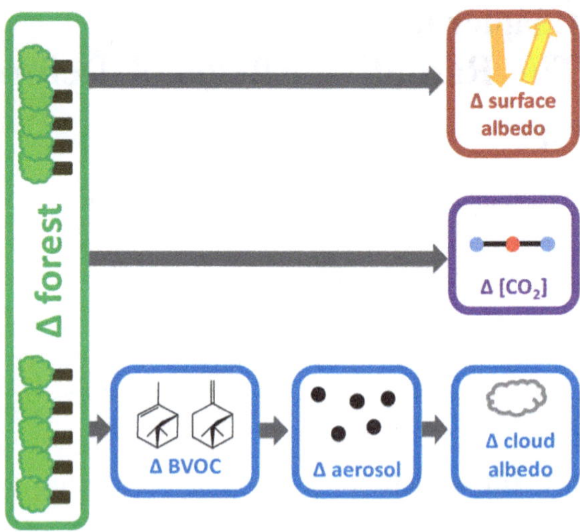


**Fig. 6.1** Summary of the impacts of deforestation that are examined in this chapter

For this work the CLM was used in the offline configuration, i.e., not coupled to either the CAM or CCSM, and atmospheric forcing (precipitation, solar radiation and atmospheric pressure, specific humidity, temperature and wind) was taken from the observationally derived dataset of Qian et al. [15], based on NCEP/NCAR reanalysis. All simulations are performed for the year 2000. Since the simulations were run without an interactive carbon-nitrogen cycle, a 40 year spin-up period was sufficient to allow the soil moisture of the CLM to establish equilibrium with the driving meteorology.

The surface of each grid cell in the CLM is divided into different plant functional types (PFTs; Table 6.1), plus non-vegetated surface. Figure 6.2 indicates the dominant PFT in each grid cell; the distribution of PFTs, and their associated leaf area indices (LAI; $m^2_{leaf\ area}\ m^{-2}_{ground\ area}$), are obtained from MODIS (MODerate Resolution Imaging Spectroradiometer) satellite data [9, 11]. This standard distribution of PFTs and LAIs will form the basis of the control, i.e., no deforestation, scenario.

#### 6.2.1.1 MEGANv2.1

The flux $F$ ($\mu g\ m^{-2}_{ground\ area}\ h^{-1}$) of a BVOC, $i$, to the atmosphere is calculated as in Eq. 6.1. Here, $\gamma_i$ is an emission activity factor accounting for responses to meteorological and phonological conditions and $\varepsilon_{i,j}$ is the PFT specific emission factor (given in Table 6.1) at standard conditions of light (photosynthetically active radiation flux of 1,000 μmol $m^{-2}$ $s^{-1}$), leaf temperature (303.15 K) and leaf area (for each vegetation type $j$), with fractional grid cell coverage of $\chi_j$.

**Table 6.1** Plant functional types, land area covered by each in the control experiment, and their emissions factors for isoprene and the dominant monoterpene (α-pinene); taken from Guenther et al. [5]

| PFT No. (Fig. 6.2) | Plant functional type | Land area covered ($10^{12}$ km$^2$) | Emission factor, ε (μg m$^{-2}$ h$^{-1}$) | | Replaced with in deforestation scenarios |
|---|---|---|---|---|---|
| | | | Isoprene | α-pinene | |
| 1 | Needleleaf evergreen tree: temperate | 5.46 | 600 | 500 | $C_3$ grass |
| 2 | Needleleaf evergreen tree: boreal | 10.6 | 3,000 | 500 | $C_3$ Arctic grass |
| 3 | Needleleaf deciduous tree: boreal | 6.46 | 1 | 510 | $C_3$ Arctic grass |
| 4 | Broadleaf evergreen tree: tropical | 15.6 | 7,000 | 600 | $C_4$ grass |
| 5 | Broadleaf evergreen tree: temperate | 2.64 | 10,000 | 400 | $C_3$ grass |
| 6 | Broadleaf deciduous tree: tropical | 12.9 | 7,000 | 600 | $C_4$ grass |
| 7 | Broadleaf deciduous tree: temperate | 5.33 | 10,000 | 400 | $C_3$ grass |
| 8 | Broadleaf deciduous tree: boreal | 2.14 | 11,000 | 400 | $C_3$ Arctic grass |
| | *PFTs that are not modified* | | | | |
| 9 | Evergreen temperate shrub | 0.18 | 2,000 | 200 | – |
| 10 | Deciduous temperate shrub | 4.15 | 4,000 | 300 | – |
| 11 | Deciduous boreal shrub | 9.33 | 4,000 | 200 | – |
| 12 | $C_3$ Arctic grass | 4.94 | 1,600 | 2 | – |
| 13 | $C_3$ grass | 14.3 | 800 | 2 | – |
| 14 | $C_4$ grass | 13.2 | 200 | 2 | – |
| 15 | Crop | 16.3 | 1 | 2 | – |

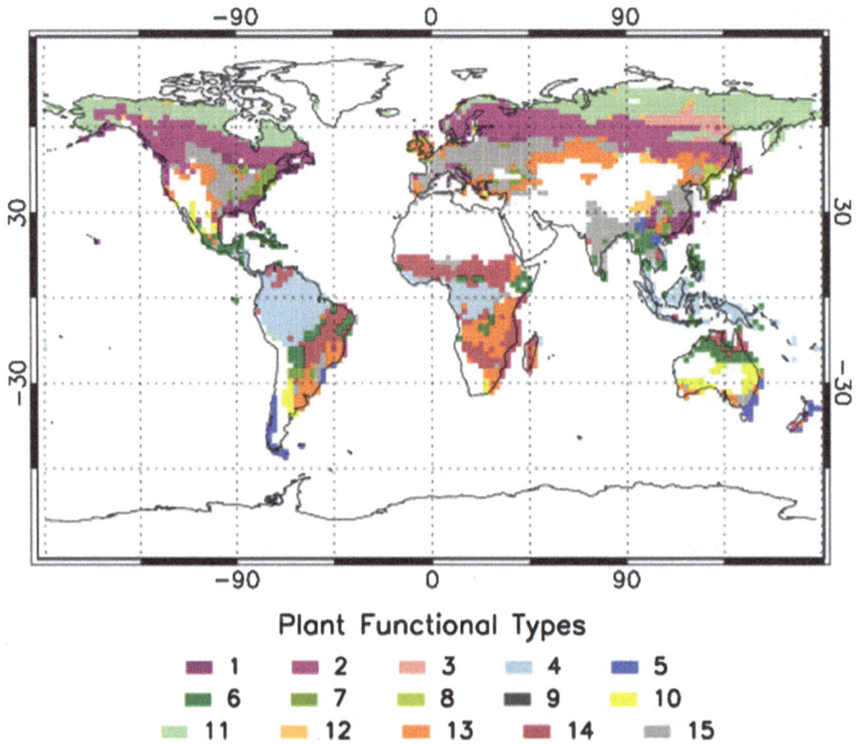


**Fig. 6.2** Dominant PFT (i.e., highest fraction occupied) in each grid cell for the standard land-cover configuration of the CLM; legend values correspond to Table 6.1

$$F_i = \gamma_i \sum_j \varepsilon_{i,j} \chi_j \tag{6.1}$$

Each emission activity factor, $\gamma_i$, is calculated according to Eq. 6.2, where $C_{CE}$ represents the canopy environment constant (derived such that emission activity will equal 1 under standard conditions), and $\gamma_P$, $\gamma_T$, $\gamma_A$, $\gamma_{SM}$ and $\gamma_C$ are scaling terms for light, temperature, leaf age, soil moisture and $CO_2$ inhibition (for isoprene only; Sect. 1.2.4.1), respectively; expressions for $\gamma$ terms are given in Sect. 2.2 of Guenther et al. [5].

$$\gamma_i = C_{CE} LAI \gamma_P \gamma_T \gamma_A \gamma_{SM} \gamma_C \tag{6.2}$$

Figure 6.3 indicates the fraction of the global total monoterpene emission, simulated by MEGANv2.1, contributed by individual compounds. Throughout the year, α-pinene comprises the largest fraction of the emission at almost 50 %, followed by trans-β-ocimene and β-pinene at between 10 and 15 %; consistent with the global emission estimates in Table 6.1. This confirms that, as a

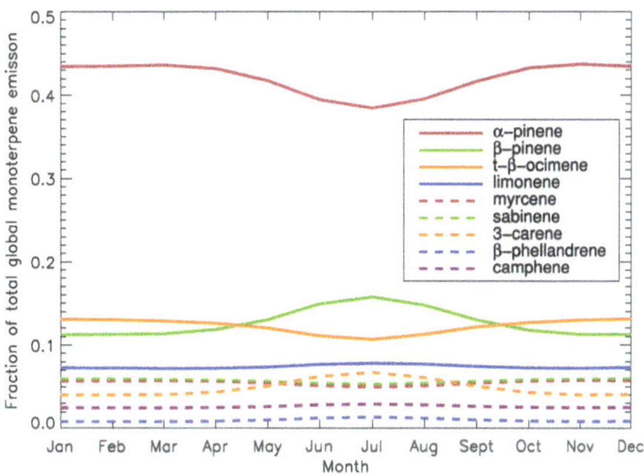


**Fig. 6.3** Fraction of global total monoterpene emission contributed by individual compounds (as simulated by MEGANv2.1); only those compounds making a greater contribution than 1 % are shown

simplification in GLOMAP, applying the reaction characteristics of α-pinene to all monoterpenes is justifiable.

#### 6.2.1.2 Comparison to Previous BVOC Emission Inventory

Using the control distribution of PFTs (Fig. 6.2), the CLM generates 480 Tg(C) $a^{-1}$ of isoprene, slightly less than the 503 Tg(C) $a^{-1}$ in the GEIA inventory, and a total monoterpene emission of 140 Tg(C) $a^{-1}$, slightly more than the 127 Tg(C) $a^{-1}$ in the GEIA inventory. Figure 6.4 shows the ratio of BVOC emissions in the GEIA inventory to those generated by the CLM. Monoterpenes and isoprene show a similar spatial pattern in the relative magnitude of emission between the two datasets. At high northern latitudes, BVOC emissions generated by the CLM (of monoterpenes in particular) during the summertime are more than a factor of 2 lower than those taken from the GEIA inventory (Fig. 6.4, upper right). Conversely, CLM emissions simulated for Central Africa and the Amazon are a factor of 2 higher than those in the GEIA inventory. This pattern is consistent with the findings of Guenther et al. [5]. The GEIA inventory was derived using the original Guenther et al. [4] algorithms, which have been updated for use in MEGAN and combined with new land-cover data and different meteorology; consequently, the reasons for the differences between the two datasets are not immediately clear.

To test the BVOC emissions generated using the CLM, the seasonal cycle in atmospheric monoterpene concentration at Hyytiälä, Finland (24.3°E, 61.9°N), simulated by GLOMAP-mode using the CLM emissions was compared to that simulated using the GEIA inventory (see Sect. 2.1.2.2), and the observations of Lappalainen et al. [7] and Hakola et al. [6]. As shown in Fig. 6.5, using the CLM emissions results in an underestimation of the summertime peak in monoterpene concentration observed at Hyytiälä, which is better captured when the GEIA inventory is used.

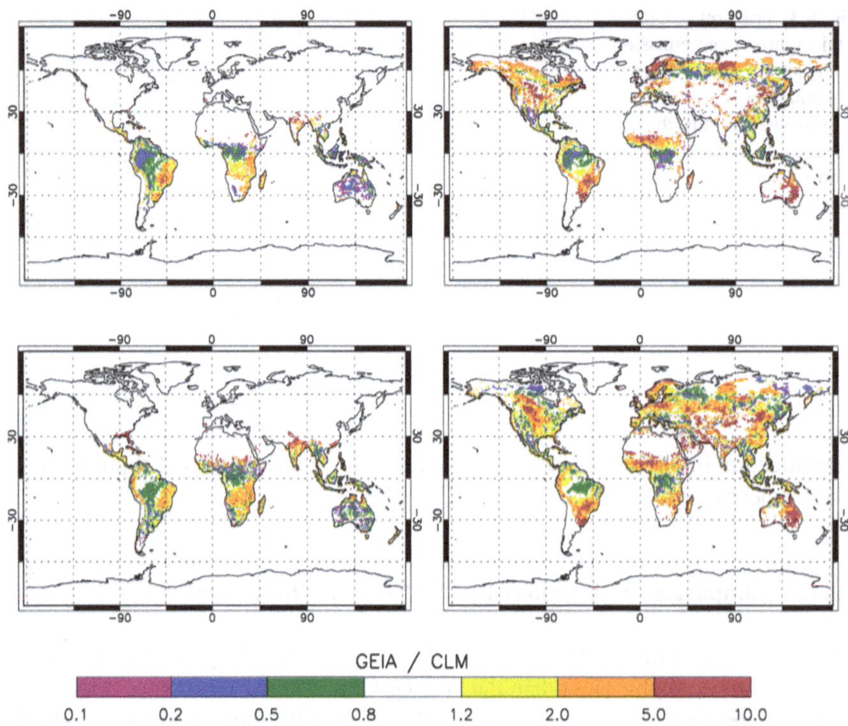


**Fig. 6.4** Emission ratio (GEIA ÷ CLM) for total monoterpenes (*upper*) and isoprene (*lower*) during January (*left*) and July (*right*)

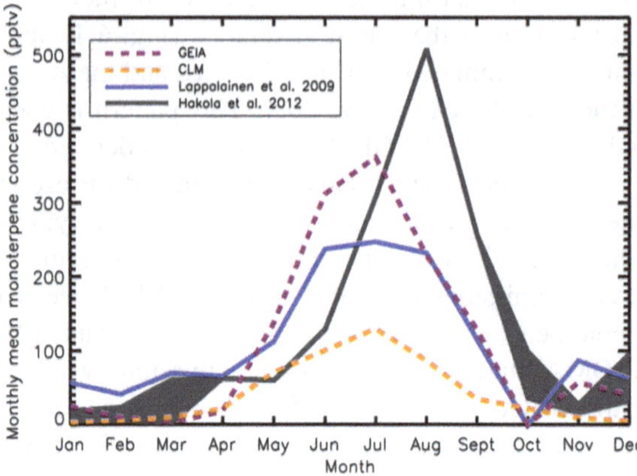


**Fig. 6.5** Monthly mean monoterpene concentrations; observed (*grey* in 2011 by Hakola et al. [6], *blue* in 2006–2007 by Lappalainen et al. [7]) and simulated by GLOMAP-mode using the GEIA emission inventory (*purple dotted*) and emissions generated by the CLM (*orange dotted*)

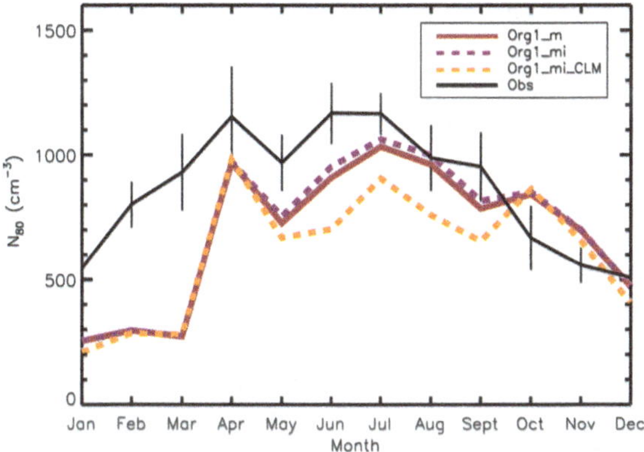


**Fig. 6.6** Multi-annual monthly mean observed (*black line*) and simulated (*Orgl_m* (*red*), *Orgl_mi* (*purple dashed*) and *Orgl_mi_CLM* (*orange dashed*)) seasonal cycle in $N_{80}$ concentration at Hyytiälä

## *6.2.2 Changes to GLOMAP-Mode Model Setup*

The simulations in this chapter use the Org1 new particle formation scheme (Eq. 3.2). In the standard configuration of GLOMAP-mode, one gas-phase tracer is used to represent the oxidation products of monoterpenes and isoprene. Consequently, for the simulations that included organically mediated nucleation in Chap. 3 (e.g., *Orgl_m*), only monoterpene emissions were included to prevent the oxidation products of isoprene contributing to new particle formation (see Sects. 1.2.4.4 and 3.2.2). Here, an additional gas-phase tracer is added to GLOMAP-mode so that the products of monoterpene and isoprene oxidation may be tracked independently. The product of monoterpene oxidation contributes to both new particle formation and condensational growth, whilst the product of isoprene oxidation contributes only to condensational growth.

The new model configuration (*Orgl_mi*) was tested using the original GEIA inventory of BVOC emissions; simulated $N_{80}$ concentrations were compared to multi-annual observations at Hyytiälä and the original *Orgl_m* configuration used in Chap. 3 (Fig. 6.6). The Pearson correlation coefficient, *R*, is virtually unchanged (0.62 with the *Orgl_mi* configuration, as compared to 0.61 using *Orgl_m* (Sect. 3.4.1)).

When the *Orgl_mi* configuration is combined with BVOC emissions generated by the CLM (*Org_mi_CLM*), *R* at Hyytiälä is reduced to 0.55, mainly due to a decrease in $N_{80}$ concentrations between May and September (Fig. 6.6). This suggests that, at Hyytiälä at least, GLOMAP does not generate an appropriate amount of SOA when BVOC emissions from the CLM are used; however, uncertainties in the formation processes of SOA prevent any conclusive statement.

For the experiments performed here, the emissions generated by the CLM and MEGANv2.1 are retained in order to be consistent with the land-use changes being examined, however, the differences between the GEIA and CLM/MEGANv2.1 datasets warrant further investigation.

### *6.2.3 Deforestation Experiments*

To quantify the radiative impact of a reduction in biogenic SOA due to deforestation, several idealised deforestation simulations were performed with the CLM. A control simulation was performed in which no modifications were made to the distribution of PFTs. For each deforestation scenario, the forested fraction of each grid cell in the deforested region was replaced with a relevant grass PFT (detailed in Table 6.1), according to the regions shown in Fig. 6.7, i.e. global (90°N–90°S), boreal (90°–50°N), temperate (50°–20°N and 20°S–50°N) and tropical (20°N–20°S). To avoid scaling up potentially inaccurate LAIs, derived from satellite observations of a small initial area of PFT, the LAIs for the grass PFTs (used to replace trees) were updated with the relevant latitudinal averages.

The CLM was used to generate emissions of monoterpenes (speciated emissions are combined) and isoprene for each deforestation scenario, following the 40-year spin-up period, after which soil moisture and BVOC emission levels were seen to stabilise to constant values. The BVOC emissions were then passed to GLOMAP-mode in order to simulate the perturbed aerosol distribution.

The dry deposition of aerosol is affected by land surface-type, with deposition velocities ($Vel_d$) generally higher over trees than grass (e.g., [14, 17]). The representation of dry deposition in GLOMAP is described in Sect. 2.1.4.6, where the surface roughness length ($z_0$), and collection efficiencies (Brownian diffusion; $E_b$, impaction; $E_{im}$ and interception; $E_{in}$) are dependent upon surface type. Across the deforested region, the surface characteristics are modified from those of trees to grass (Table 6.2) according to Zhang et al. [17].

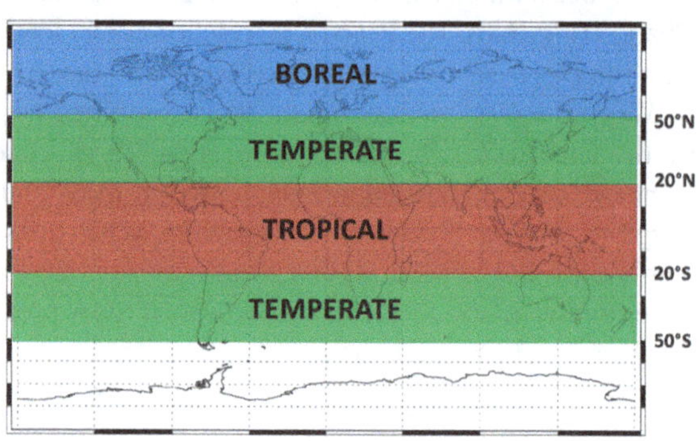


**Fig. 6.7** Regions over which forest is replaced in each deforestation experiment

**Table 6.2** Surface type characteristics used in GLOMAP-mode for forest and grass

| | Forest | Grass |
|---|---|---|
| Roughness length ($z_0$) | 0.1–3 m | 0.1 m |
| Characteristic radius (A) | 5 mm | 2 mm |

The standard wildfire emissions (Sect. 2.1.3) are maintained in the deforested regions, to represent fires potentially required to maintain an absence of forest on what would be naturally forested land.

#### 6.2.3.1 Radiative Effects of Deforestation

To estimate the radiative impacts described in Fig. 6.1, several calculations were performed using the E-S offline radiative transfer model.

Firstly, changes to CDNC (calculated offline from GLOMAP-mode output) associated with the deforestation scenarios were determined, as described in Chap. 4, and the E-S model was used to evaluate the first AIE resulting from the perturbation to cloud droplet effective radii. The cloud fraction fields (taken from ISCCP for the year 2000) were held constant between simulations.

Secondly, monthly mean surface shortwave albedos were calculated for each deforestation scenario as the ratio of the amount of shortwave radiation reflected by ($SW_R$) and incident upon ($SW_I$) each grid cell, as calculated by the CLM. In the CLM, LAIs are adjusted for burial by snow, according to snow depth and vegetation height, which in turn affects $SW_R$. The E-S offline radiative transfer model was then used to evaluate the radiative impact of the change in albedo between the various scenarios by comparing the net (SW + LW) top-of-atmosphere flux.

Lastly, the E-S model was used to estimate the radiative impact of changes to $CO_2$ concentration associated with deforestation. A coupled climate carbon-cycle model would be required to comprehensively evaluate the impact of deforestation on atmospheric $CO_2$ concentration. In its absence, changes to $CO_2$ concentration were obtained from Bala et al. [1] in which an integrated carbon-climate model is used to assess various impacts of simulated forest removal in the year 2000. When forests are replaced by grass in these simulations, the carbon they stored is gradually added to the litter pool. The simulations proceed for 100 years following deforestation, allowing the carbon released (a total of 818 Pg(C) over 100 years) to partition amongst the atmosphere, ocean and land carbon sinks. Table 6.3 details the atmospheric $CO_2$ concentration for each deforestation scenario; the regional boundaries are the same as Fig. 6.7. Anthropogenic $CO_2$ emissions follow the SRES A2 scenario from 2000 to 2100 [12], resulting in a control $CO_2$ concentration of 732 ppm in 2100. The E-S model was then used to evaluate the radiative impact of the change in $CO_2$ concentration, between each deforestation scenario and the control scenario, by comparing the net (SW + LW) fluxes at the tropopause, and correcting for stratospheric temperature re-adjustment according to Myhre et al. [10], by revising the calculated values downward by 15 %.

**Table 6.3** Atmospheric $CO_2$ concentrations after simulated deforestation; taken from Bala et al. [1]

| Simulation | $CO_2$ concentration in 2100 (ppm) |
|---|---|
| Control | 732 |
| Global deforestation | 1,113 |
| Boreal deforestation | 737 |
| Temperate deforestation | 842 |
| Tropical deforestation | 1,031 |

Finally, the combined radiative effect was evaluated by modifying the surface albedo, $CO_2$ concentration and cloud droplet effective radius simultaneously, for each deforestation scenario, in the E-S model.

These simulations do not account for changes in evapotranspiration following deforestation, which would likely decrease, resulting in a positive radiative effect. Nor do they account for the potential interaction of the radiative effects examined.

## 6.3 Results

Table 6.4 reports global BVOC emission totals, and amount of SOA generated, for the control simulation and each regional deforestation scenario. Globally replacing forests with grass reduces the global total isoprene emission by approximately 87 %, to 60 Tg(C) $a^{-1}$, and the global total monoterpene emission by approximately 94 % to 8 Tg(C) $a^{-1}$. As a result, simulated SOA production is reduced by 91 %.

Most of this reduction in emission is due to the removal of trees at tropical latitudes (20°N–20°S), which reduces the isoprene and monoterpene emission totals by 72 and 74% respectively. Despite the large areas of forests north of 50°N, simulated boreal deforestation reduces global isoprene and monoterpene emissions by only 1 and 5% respectively. However, as shown in Sect. 6.1.2.1, the CLM may be underestimating high latitude monoterpene emissions.

### 6.3.1 *First Aerosol Indirect Effect*

Table 6.5 reports the change to global annual mean CDNC for each deforestation scenario, when compared to the control simulation. Global deforestation, which leaves only 3 Tg(SOA) $a^{-1}$, reduces the global annual mean CDNC by 7.6 %. This reduction in CDNC yields a global annual mean first AIE of +0.26 W $m^{-2}$ with the spatial pattern shown in Fig. 6.8 (upper left).

For a number of reasons, the positive first AIE for global deforestation is smaller in magnitude than the equivalent negative AIE calculated for monoterpene

**Table 6.4** BVOC emissions and amount of SOA generated during deforestation experiments

| | Global annual total | | | | | |
|---|---|---|---|---|---|---|
| | Isoprene emission (Tg(C) $a^{-1}$) and percentage change from control | | Total monoterpene emission (Tg(C) $a^{-1}$) and percentage change from control | | SOA generated (Tg(SOA) $a^{-1}$) and percentage change from control | |
| Control | 480 | – | 140 | – | 35.6 | – |
| Global deforestation | 60 | −87 % | 8 | −94% | 3.1 | −91 % |
| Boreal deforestation | 475 | −1 % | 133 | −5% | 34.4 | −3 % |
| Temperate deforestation | 412 | −14 % | 119 | −15% | 30.1 | −15 % |
| Tropical deforestation | 133 | −72 % | 36 | −74% | 9.7 | −73 % |

**Table 6.5** Summary of results from deforestation scenarios

| | Percentage change to global annual mean CDNC | AIE (W $m^{-2}$) | ΔSOA (Tg(SOA)) | AIE per change in SOA [AIE/ΔSOA] (W $m^{-2}$ Tg(SOA)$^{-1}$) |
|---|---|---|---|---|
| Global deforestation | −7.6 | +0.26 | 32.5 | $0.79 \times 10^{-2}$ |
| Boreal deforestation | −1.4 | +0.02 | 1.2 | $1.49 \times 10^{-2}$ |
| Temperate deforestation | −2.7 | +0.09 | 5.5 | $1.68 \times 10^{-2}$ |
| Tropical deforestation | −2.5 | +0.12 | 25.8 | $0.47 \times 10^{-2}$ |

SOA in Chap. 4 (*Org1_m*; −0.77 W $m^{-2}$). Firstly, since the source of organic oxidation products is not completely removed when forests are replaced with grass, nucleation is not completely suppressed, as it was in the absence of biogenic SOA in Chaps. 3 and 4. Secondly, the BVOC emissions generated by the CLM are lower than those from the GEIA inventory in some of the more pristine regions (e.g., high northern latitudes, Australia and the south-east coast of South America) that were important for the first AIE simulated in Chap. 4.

Tropical deforestation generates the largest AIE (+0.12 W $m^{-2}$) of the individual regions due to substantial year-round decreases in CDNC (up to 50 %) and the abundance of low-level clouds over tropical regions. The first AIE calculated due to tropical deforestation will be sensitive to the assumptions made regarding wildfire and biomass burning emissions, since these hydrophobic particles may be *physically aged* by SOA and subsequently act as CCN. Here, the standard GFED emission set for biomass burning is retained, but a more realistic scenario could include gradual deforestation (i.e., gradual reduction in the production of biogenic SOA) and shifting patterns of deforestation fire emissions.

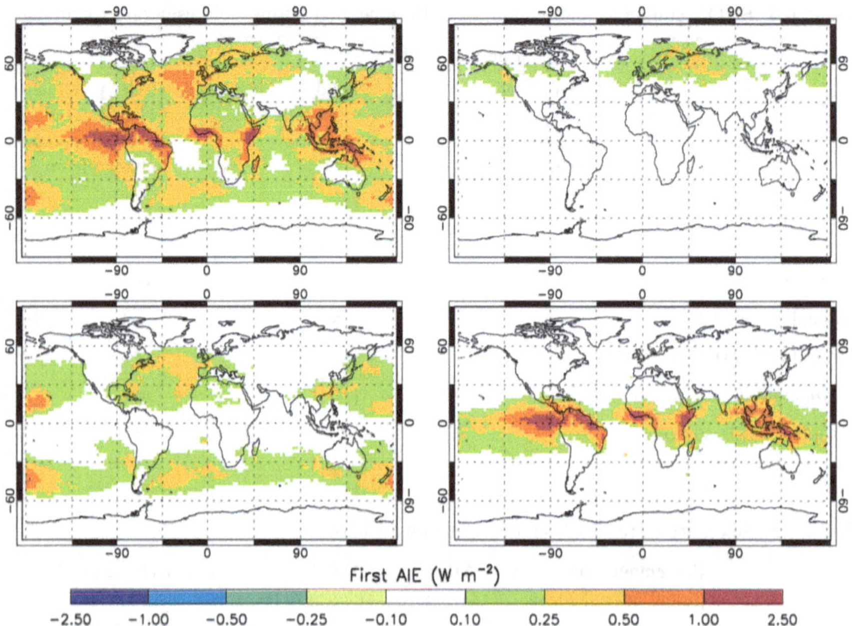


**Fig. 6.8** Annual mean first AIE (W m$^{-2}$) resulting from reduction in SOA due to replacement of global (*upper left*), boreal (*upper right*), temperate (*lower left*) and tropical (*lower right*) forests with grass

Temperate deforestation gives the largest AIE per change in SOA ($1.68 \times 10^{-2}$ W m$^{-2}$ Tg(SOA)$^{-1}$). This occurs because the removal of biogenic SOA in temperate regions affects CDNC across the remote northern hemisphere (NH) and southern hemisphere (SH) ocean regions, with the influence of the change spreading into regions of very high cloud fraction (Fig. 6.8, lower left).

Boreal deforestation reduces the global annual mean CDNC by only 1.4 %, but regional reductions over northern Russia and Canada, evident in the first AIE (Fig. 6.8, upper right), in the NH summertime exceed 35 %. Figure 6.9 shows the seasonal cycle in regional mean AIE over each deforested region, highlighting the summertime peak in first AIE of approximately +0.4 W m$^{-2}$ over the boreal region. Regionally, the combined contribution of temperate and boreal deforestation leads to a summertime (JJA mean) local first AIE between +0.3 W m$^{-2}$ and +1.5 W m$^{-2}$ across much of the region between 40 and 80°N (Fig. 6.10).

### 6.3.2 Other Radiative Effects of Forests

To assess the relative importance of the calculated first AIE from deforestation, the radiative effects (RE) of changes to surface albedo and atmospheric $CO_2$

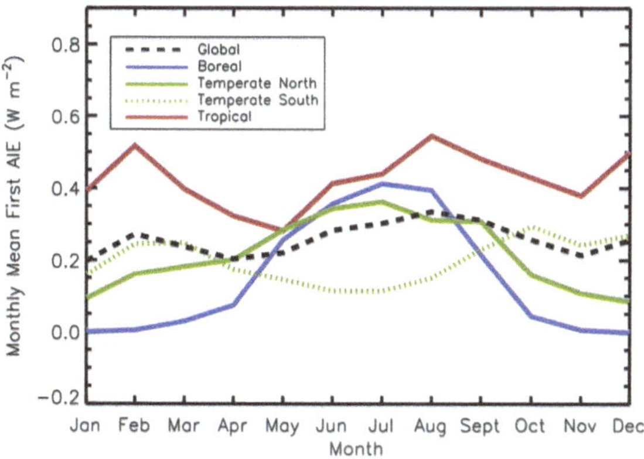


**Fig. 6.9** Monthly-mean first AIE (W m$^{-2}$), over the boreal (*blue*), NH temperate (*full green*), SH temperate (*green dotted line*) and tropical (*red*) regions, during the global deforestation simulation

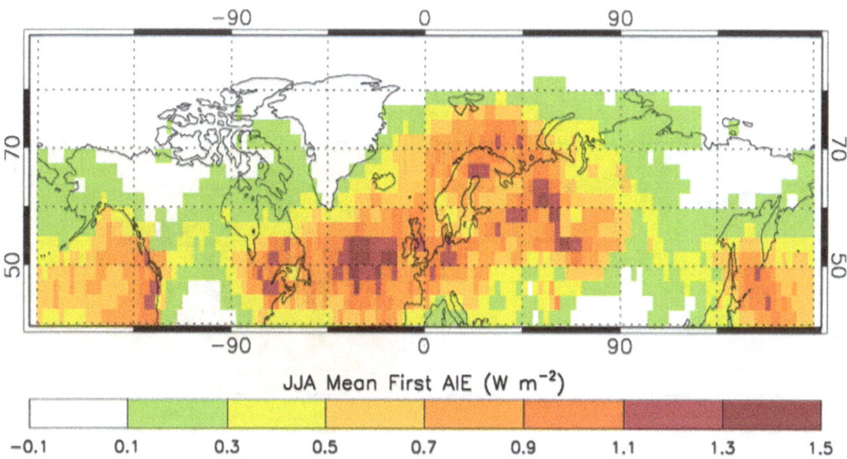


**Fig. 6.10** NH summertime (June–July–August) mean first AIE (W m$^{-2}$) due to global deforestation

concentration were also calculated. Table 6.6 reports the global annual mean REs for each deforestation scenario.

The global annul mean RE due to surface albedo change (−0.96 W m$^{-2}$) is dominated by boreal deforestation, which alone exerts a global annual mean RE of −0.51 W m$^{-2}$. Over Canada and Russia, the increase in surface albedo, and enhanced snow cover, due to deforestation leads to regional REs up to −25 W m$^{-2}$ (Fig. 6.11, left), consistent with Betts [2]. Snow cover and the level of incident solar radiation combine to give a peak RE from surface albedo change during March and April (Fig. 6.12) over the boreal region.

**Table 6.6** Summary of global annual mean radiative effects due to change in BVOC emission, surface albedo and atmospheric $CO_2$ concentration (adjusted) associated with each deforestation scenario

| | Global annual mean radiative effects | | | | |
|---|---|---|---|---|---|
| | AIE ($W\ m^{-2}$) | RE due to Δalbedo ($W\ m^{-2}$) | RE due to $\Delta[CO_2]^a$ ($W\ m^{-2}$) | RE due to Δalbedo plus $\Delta[CO_2]^b$ ($W\ m^{-2}$) | Total RE[b] ($W\ m^{-2}$) |
| Global deforestation | +0.26 | −0.96 | +2.20 | +1.24 | +1.50 |
| Boreal deforestation | +0.02 | −0.51 | +0.03 | −0.48 | −0.46 |
| Temperate deforestation | +0.09 | −0.27 | +0.75 | +0.48 | +0.57 |
| Tropical deforestation | +0.12 | −0.18 | +1.80 | +1.62 | +1.74 |

[a] Based on $CO_2$ concentrations 100 years after deforestation from Bala et al. [1] (Table 6.3) and revised downwards by 15 % to account for stratospheric temperature readjustment, following Myhre et al. [10]

[b] Sum of radiative effects when radiative transfer model run individually for each perturbation

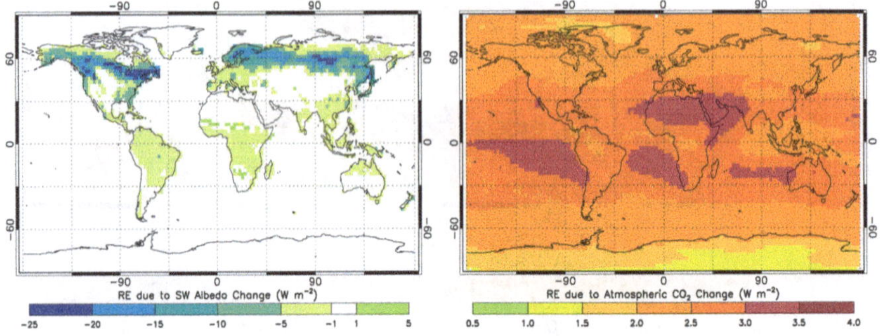


**Fig. 6.11** Annual mean radiative effect of change in surface albedo (*left*) and atmospheric $CO_2$ concentration (non-adjusted values; *right*) due to simulated global deforestation

The radiative effect from $CO_2$ (Fig. 6.11, right) follows a logarithmic relationship with the ratio of atmospheric $CO_2$ concentration in the deforestation scenario to that in the control. In Bala et al. [1], global deforestation increases the atmospheric $CO_2$ concentration by 381 ppm after 100 years (Table 6.3), giving an RE of +2.20 $W\ m^{-2}$. In reality, the radiative effect from $CO_2$ would be sensitive to the mode of deforestation, the fate of the carbon stored by the trees and the timescale over which it is released.

The radiative effects combine linearly (tested using the non-adjusted $CO_2$ RE values but not shown here), providing similar values when the radiative transfer model is run with separate perturbations to cloud droplet effective radii, surface

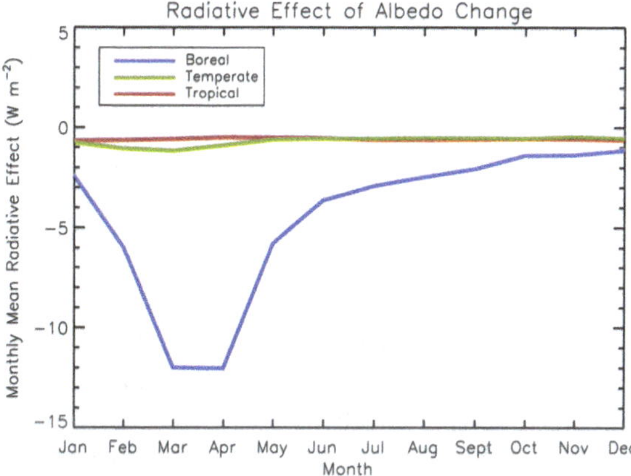


**Fig. 6.12** Seasonal cycle in regional mean RE due to surface albedo change only

albedo and $CO_2$ concentration, and when the individual radiative effects are summed. In the absence of the first AIE, the additive RE from the change to $CO_2$ concentration and surface albedo due to global deforestation would be +1.24 W $m^{-2}$; this is increased by 21 % to +1.50 W $m^{-2}$ when the first AIE is included (Table 6.6).

In order to make a first estimate of the potential first AIE due to forest removal, these experiments make several assumptions. Firstly, a new particle formation rate that is dependent on the presence of biogenic organic compounds is used in GLOMAP-mode; an approach which is supported by results in Chap. 3. Secondly, the removal of all forests in one instance is not intended to represent a realistic course of events, but to allow an estimation of the radiative implications of an absence of forests. Accordingly, the radiative effects of actual deforestation could be different, i.e., a gradual reduction in SOA production, combined with increased forest fire emissions. Additionally, these experiments do not include the impact of changes to evapotranspiration, a reduction in which (due to forest removal) would likely contribute an additional warming (e.g., [1, 3]). This would require a coupled hydrological cycle which is beyond the scope of the study presented here, but an avenue for further investigation.

## 6.4 Summary and Conclusions

In this chapter, a land-surface model, aerosol microphysics model, and radiative transfer model were combined (offline) to assess the climatic impact of the idealised deforestation of various regions.

The first AIE due to a reduction in SOA production associated with deforestation has been estimated for the first time. Globally, the first AIE due to the replacement of forests with grass (+0.26 W $m^{-2}$) increases the positive radiative effect (net effect from increased $CO_2$ concentration and surface albedo) of deforestation by 21 %, suggesting that present-day deforestation may be warming the climate more than previously thought. In further work, this model framework will be used to assess the impact of historical deforestation, i.e., in a pre-industrial atmosphere.

Of the three regional deforestation scenarios examined, tropical deforestation resulted in the largest absolute first AIE (+0.12 W $m^{-2}$), increasing the net effect by 7 %. Since the AIE is strongest in the tropics, the latitudinal (i.e., from 0° to 90°N) gradient in the total radiative effect from forest removal is increased. The first AIE calculated for tropical deforestation will be particularly sensitive to the assumptions made regarding wildfire and deforestation fire emissions. In order to calculate a more realistic first AIE for tropical deforestation, this approach could be improved by gradually reducing the forested area (i.e., when calculating the changing BVOC emission), and simultaneously increasing emissions from deforestation fires (used in GLOMAP-mode), over a series of simulations.

Previous studies suggested that the strength of the albedo effect gives simulated boreal deforestation an overall cooling effect (e.g., [1, 3]). The present analysis does not change that conclusion, however, the first AIE for boreal deforestation (+0.02 W $m^{-2}$) calculated here may be an underestimate if the BVOC emissions generated by the CLM are too low at high northern latitudes.

Deforestation in temperate regions was found to give the largest first AIE per change in SOA, increasing the combined RE by 19 %. Snyder et al. [16] found that the temperature change induced by NH temperate deforestation was seasonally dependent, with a simulated warming between June and November, but a cooling between December and May. The present analysis suggests that inclusion of the first AIE due to reduced SOA production would enhance this summertime warming (Fig. 6.10), and by imposing an additional positive RE may increase the fraction of the year for which NH temperate deforestation would induce a warming (Fig. 6.9).

Most present-day afforestation is occurring in temperate regions; the present analysis suggests that this may have been exerting more of a cooling effect on the climate than would have been attributed to $CO_2$ sequestration alone. Further work will explore the potential negative RE from increasing forest cover in particular temperate regions (e.g., China), incorporating species-specific BVOC emissions.

The first AIE calculated in this chapter will be sensitive to the assumptions made when calculating the impact of reduced biogenic SOA on particle concentrations, in particular whether or not BVOC oxidation products are involved in the initial stages of new particle formation. It was shown in Chap. 3 that this is likely to be the case, however our understanding of the mechanisms that drive new particle formation in the atmosphere remains incomplete; subsequent estimates of this effect will therefore be sensitive to any further insights into this process.

## References

1. Bala G et al (2007) Combined climate and carbon-cycle effects of large-scale deforestation. PNAS 104(16):6550–6555
2. Betts RA (2000) Offset of the potential carbon sink from boreal forestation by decreases in surface albedo. Nature 408(6809):187–190
3. Davin EL, de Noblet-Ducoudré N (2010) Climatic impact of global-scale deforestation: radiative versus nonradiative processes. J Clim 23(1):97–112
4. Guenther A et al (1995) A global model of natural volatile organic compound emissions. J. Geophys. Res. 100(D5):8873–8892
5. Guenther AB et al (2012) The model of emissions of gases and aerosols from nature version 2.1 (MEGAN2.1): an extended and updated framework for modeling biogenic emissions. Geosci Model Dev 5(6):1471–1492
6. Hakola H et al (2012) In situ measurements of volatile organic compounds in a boreal forest. Atmos Chem Phys 12(23):11665–11678
7. Lappalainen HK et al (2009) Day-time concentrations of biogenic volatile organic compounds in a boreal forest canopy and their relation to environmental and biological factors. Atmos Chem Phys 9(15):5447–5459
8. Lawrence DM et al (2011) Parameterization improvements and functional and structural advances in Version 4 of the Community Land Model. J Adv Model Earth Syst 3(1):M03001
9. Lawrence PJ, Chase TN (2007) Representing a new MODIS consistent land surface in the community land model (CLM 3.0). J Geophys Res Biogeosci 112(G1):G01023
10. Myhre G et al (1998) New estimates of radiative forcing due to well mixed greenhouse gases. Geophys Res Lett 25:2715–2718
11. Myneni RB et al (2002) Global products of vegetation leaf area and fraction absorbed PAR from year one of MODIS data. Remote Sens Environ 83(1–2):214–231
12. Nakicenovic N et al (2000) Special report on emission scenarios. Nakicenovic N, Swart R
13. Oleson KW et al (2010) Technical description of version 4.0 of the community land model (CLM). NCAR Technical Note NCAR/TN-478+STR. Boulder, Colorado, National Centre for Atmospheric Research: 257
14. Petroff A et al (2008) Aerosol dry deposition on vegetative canopies. Part I: review of present knowledge. Atmos Environ 42(16):3625–3653
15. Qian T et al (2006) Simulation of global land surface conditions from 1948 to 2004. Part I: forcing data and evaluations. J Hydrometeorol 7(5):953–975
16. Snyder PK et al (2004) Evaluating the influence of different vegetation biomes on the global climate. Clim Dyn 23:279–302
17. Zhang L et al (2001) A size-segregated particle dry deposition scheme for an atmospheric aerosol module. Atmos Environ 35(3):549–560

# Chapter 7
# Conclusions, Implications and Further Work

## 7.1 Summary of Results

This thesis has explored the role of biogenic secondary organic aerosol (SOA) in the present-day, and pre-industrial, atmosphere. Chapters 3–5 examined the behaviour of SOA, and the radiative impact of its presence, whilst Chap. 6 focussed on the climatic significance of forest-derived SOA.

In Chap. 3, a global aerosol microphysics model (GLOMAP-mode) was used to quantify changes to total particle (greater than 3 nm dry diameter; $N_3$) and cloud condensation nuclei (CCN) number concentration. It was shown that the impact of biogenic SOA on the aerosol distribution is complex, and sensitive to many uncertain parameters and processes.

The effect of biogenic SOA on $N_3$ concentration is dependent upon the mechanism used to simulate new particle formation. In the absence of organically mediated new particle formation, the global annual mean $N_3$ concentration decreases in the presence of biogenic SOA (by as much as 17.5 %), due to the enhanced condensation and coagulational sinks. If organically mediated new particle formation is included, the global annual mean $N_3$ increases by as much as 142 %.

The seasonal cycle in total particle concentration (varying between $N_3$ and $N_{14}$), across sites in the continental northern hemisphere was best captured when organically mediated new particle formation was included in the model (Sect. 3.4.2). Introducing a dependency on biogenic species increases summertime total particle concentrations, relative to wintertime, and improves the correlation coefficient from 0.23 when new particle formation is dependent only on the concentration of sulphuric acid, to 0.40 when the new particle formation rate derived in the CLOUD chamber is used.

The inclusion of biogenic SOA also improved the simulated seasonal cycle in $N_{80}$ (particles greater than 80 nm dry diameter) and reduced the model bias when compared against CCN observations (Sect. 3.4). The presence of biogenic SOA



C.E. Scott, *The Biogeochemical Impacts of Forests and the Implications for Climate Change Mitigation*, Springer Theses, DOI 10.1007/978-3-319-07851-9_7

increased the simulated global annual mean CCN concentration by between 3.6 % (when organic oxidation products are not able to physically age non-hydrophilic particles) and 45.2 % (when organic oxidation products contribute to new particle formation). In the absence of organically mediated new particle formation, most of the increase in CCN concentration occurs due to the physical ageing of initially non-hydrophilic particles, e.g., from wildfires. However, when monoterpene oxidation products directly affect the rate of new particle formation, CCN concentrations are strongly perturbed by the growth of newly formed particles to a CCN-active size.

The improved representation of the seasonal cycle, in both total particle concentrations and $N_{80}$, when organically mediated new particle formation is included in GLOMAP-mode, strongly suggests a role for a biogenic control on new particle formation. Accordingly, the sensitivity of CCN concentrations to the presence of biogenic SOA is high.

In Chap. 4, the radiative effects (REs) of the changes to particle number, size, and composition quantified in Chap. 3, were evaluated. It was shown that biogenic SOA very likely has a negative radiative effect in the present-day atmosphere, via both the *direct* and first *indirect* effect. The *direct* radiative effect from SOA was shown to be most sensitive to the amount of SOA produced in the simulation, varying from $-0.09$ W m$^{-2}$ for a source of 18.5 Tg(SOA) a$^{-1}$, to $-0.78$ W m$^{-2}$ for a source of 185 Tg(SOA) a$^{-1}$. An accurate assessment of the direct effect from SOA will require a refinement of the current wide range in estimates of the global SOA budget. The first aerosol *indirect* effect (AIE) due to biogenic SOA was shown to be very sensitive to the mechanism used to simulate new particle formation, ranging from $-0.05$ W m$^{-2}$ for a simulation including only binary homogenous nucleation (i.e., little new particle formation in the boundary layer) to $-0.77$ W m$^{-2}$ for a simulation including organically mediated new particle formation. At high northern latitudes, monoterpene emissions from boreal forests result in regional summertime AIEs of up to $-5$ W m$^{-2}$ over land, and $-8$ W m$^{-2}$ over the adjacent ocean regions.

The first *indirect radiative forcing* (RF) of anthropogenic emissions, since 1750, was found to be sensitive to the assumptions made concerning biogenic SOA (Sect. 4.4). The RF changes by 0.06 W m$^{-2}$ when the SOA production yield is varied by a factor of 10 and by 0.12 W m$^{-2}$ when the approach to modelling new particle formation (and the involvement of biogenic oxidation products) is changed. This highlights, as previously demonstrated by Schmidt et al. [33] for volcanic eruptions, that in order to understand the radiative effects of human activities, a comprehensive understanding of the background atmospheric state in pre-industrial times is required.

In Chap. 5 it was shown that the first AIE calculated from biogenic SOA is sensitive to the model treatment of SOA volatility. Whilst the first stage oxidation products of many BVOCs are semi-volatile, their continued oxidation in the gas- or particle-phase reduces their volatility further still and may provide a supply of very low volatility compounds to the particle-phase. Two different, but commonly implemented approaches to the partitioning of SOA amongst the existing particle distribution were examined. The first, a *kinetic* approach, distributes the SOA across the existing size distribution according to particle surface area. The second, a *thermodynamic* approach, distributes the SOA according to the existing organic mass. This thermodynamic approach suppresses the growth of newly formed particles, removing this as a source of CCN. To capture the observed growth of newly formed particles, the kinetic approach is required—however, this neglects the potential re-evaporation of semi-volatile organics, back into the gas-phase.

Accurately simulating the condensation of SOA onto the existing aerosol size distribution is important when evaluating processes that depend strongly on changes to ultrafine particle number, such as the AIE. In Chap. 5 it was shown that the first AIE from biogenic SOA is negative if the kinetic approach is taken, but positive, or negligible, if the thermodynamic approach is taken. The next step would be to combine the two approaches, with a fraction of the SOA assigned a very low volatility (theoretically, since the volatility is not actually assigned in the model) and partitioning via the kinetic route, and a fraction of the SOA assigned a moderate volatility and partitioning via the thermodynamic route. The current literature offers little constraint on the values these two fractions should take. However, since gas- and particle-phase processing tends to shift organic compounds to lower volatility (e.g., [7, 15]), one could assume that the fraction of lower volatility material should be the larger of the two.

In Chap. 6, the radiative effect of forest derived SOA in the present-day atmosphere was explored using several idealised deforestation scenarios. It was shown that including the first AIE due to reduced biogenic SOA production (+0.26 W $m^{-2}$) increased the positive radiative effect from global deforestation (i.e., the net change from increased $CO_2$ concentration and increased surface albedo) by 21 %. The magnitude of the AIE due to biogenic SOA is strongest for total tropical deforestation (+0.12 W $m^{-2}$), but the largest AIE per change in SOA was found for temperate deforestation, due to modest but widespread decreases in cloud droplet number concentration in regions of high cloud fraction.

Pongratz et al. [30] have previously estimated a RF of +0.15 W $m^{-2}$ due to historical (AD 800 to 1992) land-use change (−0.2 W $m^{-2}$ from surface albedo changes and +0.35 W $m^{-2}$ from $CO_2$ emission); results from Chap. 6 suggest that historical deforestation may have warmed the climate more than previously thought. Further work will examine the significance of the first AIE due to historical land-use change, using the dataset of Pongratz et al. [30].

Over the past decade, the rate of tropical deforestation has decreased, and the rate of temperate afforestation has increased (Sect. 1.2.1). Results from Chap. 6 would suggest that the first AIE due to changes in biogenic SOA over that time would be negative, relative to the immediately preceding period (e.g. 1960s–1990s) with high levels of tropical deforestation (i.e., that the total RF due to land-use change would be less positive than previously). If sufficiently large, this could have contributed to the observed hiatus in temperature increase observed over the last decade [22]. A further study to accurately quantify the radiative effects from recent levels of deforestation and afforestation is therefore warranted, building on the framework established in Chap. 6.

### 7.1.1 Implications for Understanding and Mitigating Future Climate Change

In order to limit equilibrium climate warming to 2 K (Sect. 1.3.2), the net RF resulting from anthropogenic activities would need to be restricted to between +2.5 and +3.9 W $m^{-2}$ (assuming an *equilibrium climate sensitivity* of between 1.9 K [24] and 3 K [20]). This calculation is based on the equilibrium temperature change, whereas the transient climate changes experienced will be sensitive to the temporal and spatial nature of forcing agents. As highlighted by Ramanathan and Xu [31], the current RF from greenhouse gas levels (approximately 3 W $m^{-2}$; Fig. 1.10) is being masked by a very uncertain negative aerosol RF which may decrease under policies designed to improve air quality.

As shown in Chap. 4, the calculated anthropogenic indirect RF (and therefore the total RF due to anthropogenic activities) is sensitive to the representation of the natural "background" state of the atmosphere, to which SOA is a significant contributor. If anthropogenic sources of aerosol reduce in the future, our understanding of the behaviour of natural components of the atmosphere will become increasingly important.

Quantifying the current global budget of biogenic SOA, and understanding how this may change in the future, should therefore be a priority for the research community. This will require an improved understanding of the behaviour of biogenic SOA, even in the present-day atmosphere. Whilst including a biogenic control on the rate of new particle formation represents a substantial improvement in our ability to model the seasonal cycle in particle number concentrations, our understanding of the process of new particle formation and growth is not complete. It is likely that different types of organic compounds are responsible for different aspects of the role of SOA in the atmosphere, e.g., formation of the initial cluster, growth of the initial clusters to an observable size, and growth of observable particles to a climatically relevant size. For example, Häkkinen et al. [14] found that the measured growth rate of 7–20 nm diameter particles showed a clear peak in the NH summertime, suggesting a biogenic control. Conversely, the observed growth rate of sub-3 nm particles did not show a seasonal cycle, suggesting that this was not driven by biogenic organic compounds, but potentially anthropogenic organic compounds, or even amines (e.g., [16, 18]; although the determination of growth rates below 3 nm is subject to large errors [37]). Accurately representing the spatial and temporal variations in the formation and growth rates of new particles will therefore require an improved understanding of the compounds that contribute at each stage.

In the future, sources of biogenic SOA will change. Emissions of BVOCs will be sensitive to changes in temperature [19, 21, 25, 35] and carbon dioxide concentrations [1, 26, 32]. The spatial distribution of BVOC emissions will also change, both through shifting land-use practises, e.g., deforestation and afforestation, biofuel growth [2, 11] and vegetation management (e.g., logging [10]), and through climate induced changes, e.g., a northwards shift of vegetation in warming climate [12], savannisation in the tropics [38], or enhanced levels of beetle infestation [3]. It was shown in Chaps. 3 and 4 that the location of emissions is important, as well as their magnitude, with emissions into relatively pristine locations able to exert a more substantial indirect radiative effect.

Reducing deforestation, and increasing afforestation, has been suggested as part of a strategy to mitigate potential future climate change (Sect. 1.3.2.1). Results from Chap. 6 suggest that deforestation, particularly in the tropics, is causing more of a positive radiative effect than previously thought and a reduction in deforestation rates should therefore be seen as a priority. One possibility to help achieve this would be the inclusion of non-carbon effects in the metrics used to assess the importance of land-use changes. In schemes such as REDD, the preservation of carbon stored on the land is considered, but it will be necessary to take the other effects of forests into consideration, e.g., their effects on rainfall [34] and the radiative effects explored in this thesis, to accurately assess the impacts of forests on the climate.

## 7.2 Further Work

Several avenues for further work have been discussed in the previous sections. Important questions to be answered are summarised briefly below, and some potential modifications to the GLOMAP-mode model are described in the following section.

- How significant is the reduction in biogenic SOA, and the associated radiative effects, attributable to historical deforestation. This could be answered using the land-use change data collated by Pongratz et al. [30] to generate BVOC emission fields (potentially using the Community Land Model) for various years (e.g., 1750, 1850, 1950, 2000), under changing climatic conditions. The impact of reduced biogenic SOA could then be quantified using GLOMAP-mode, by including anthropogenic emissions for the appropriate year.
- What is the first AIE associated with current levels of tropical deforestation, i.e., a loss of approximately 13 million hectares per year between 2000 and 2010 [8], and what are the implications of deforestation continuing at this level?
- How significant could temperate afforestation be in terms of climate change mitigation, is the negative first AIE associated with expansion of temperate forests significant? This could be answered using the framework developed in Chap. 6 by replacing grass and crop plant functional types with trees across the temperate northern hemisphere, and generating BVOC emissions using the Community Land Model, taking into consideration the particular species being planted.

### 7.2.1 Secondary Organic Aerosol in GLOMAP

The representation of SOA in GLOMAP is necessarily simple. Section 7.1 discussed the possibility of adding additional tracers to the model to reflect the range of volatilities present amongst organic oxidation products. Further to this, the behaviour of the oxidation products of compounds from the same class may even be quite different; the oxidation products of exocyclic structures (i.e., double bond outside the ring, e.g., β-pinene) tend to retain their ring structure, whilst the oxidation of endocyclic structures (i.e., double bond inside the ring, e.g., α-pinene) gives ring-opened products, yielding two classes of compounds with potentially differing behaviours in the atmosphere [4, 5]. This may be important in locations where vegetation emissions are dominated by particular compounds, for all or part of the year. Accurate speciated emissions and atmospherically relevant yields for SOA production from individual compounds would be required to improve the model in this respect, which may be possible in the future.

This thesis has not examined the role of sesquiterpenes. Whilst the estimated annual emission of sesquiterpenes (<30 Tg(C) $a^{-1}$; [9]) is far lower than either isoprene or monoterpenes, very high yields have been observed in the laboratory for SOA produced from sesquiterpene ozonolysis (e.g., [23]). Accordingly, their future inclusion in global models, potentially using emissions generated by MEGANv2.1, may be important.

### 7.2.2 Dry Deposition in GLOMAP

The representation of dry deposition in GLOMAP is relatively sophisticated, however, the categories into which the land-surface is split are limited and surface characteristics are held constant throughout the year. The current parameterisation could be updated using the approach of Petroff et al. [27, 28] and Petroff and Zhang [29] to further distinguish between broadleaf and needleleaf, evergreen and deciduous trees, and add a seasonal dependence for the surface characteristics (e.g., roughness length and collection radius).

### 7.2.3 Tropospheric Chemistry

As well as affecting the particle-phase via the formation of SOA, BVOC emissions impact the concentrations of compounds in the gas-phase, e.g., $O_3$ [13]. As such, changes to the level and distribution of BVOC emissions could affect the oxidative capacity of the atmosphere, and the lifetime of other climatically important gases, e.g., methane (e.g., [36]). The importance of these changes could be investigated using the coupled-chemistry version of GLOMAP-mode, described by Breider et al. [6] and Schmidt et al. [33].

Additionally, the application of a fixed yield for the production of SOA from BVOC oxidation does not allow for any dependency on the concentration of other atmospheric constituents (apart from the oxidants $O_3$, OH and $NO_3$), e.g., NOx, which may be important, particularly for isoprene oxidation (Sect. 1.2.4.3; [17]). An SOA production dependency on NOx concentrations could be added to the coupled-chemistry version of the model.

## References


1. Arneth A et al (2007) $CO_2$ inhibition of global terrestrial isoprene emissions: potential implications for atmospheric chemistry. Geophys Res Lett 34(18):L18813
2. Ashworth K et al (2013) Impacts of biofuel cultivation on mortality and crop yields. Nat Clim Change 3:492–496
3. Berg AR et al (2013) The impact of bark beetle infestations on monoterpene emissions and secondary organic aerosol formation in western North America. Atmos Chem Phys 13(6):3149–3161
4. Bonn B et al (2002) Influence of water vapor on the process of new particle formation during monoterpene ozonolysis. J Phys Chem A 106(12):2869–2881
5. Bonn B, Moorgat GK (2002) New particle formation during a- and b-pinene oxidation by $O_3$, OH and $NO_3$, and the influence of water vapour: particle size distribution studies. Atmos Chem Phys 2(3):183–196
6. Breider TJ et al (2010) Impact of BrO on dimethylsulfide in the remote marine boundary layer. Geophys Res Lett 37(2):L02807
7. Donahue NM et al (2011) Theoretical constraints on pure vapor-pressure driven condensation of organics to ultrafine particles. Geophys Res Lett 38(16):L16801
8. FAO (2010). Global Forest Resources Assessment (2010) FAO Forestry Paper 163. United Nations, Rome
9. Guenther AB et al (2012) The Model of Emissions of Gases and Aerosols from Nature version 2.1 (MEGAN2.1): an extended and updated framework for modeling biogenic emissions. Geosci Model Dev 5(6):1471–1492
10. Haapanala S et al (2012) Is forest management a significant source of monoterpenes into the boreal atmosphere? Biogeosciences 9(4):1291–1300
11. Hardacre CJ et al (2013) Probabilistic estimation of future emissions of isoprene and surface oxidant chemistry associated with land-use change in response to growing food needs. Atmos Chem Phys 13(11):5451–5472
12. Harsch MA et al (2009) Are treelines advancing? A global meta-analysis of treeline response to climate warming. Ecol Lett 12(10):1040–1049
13. Hewitt CN et al (2011) Ground-level ozone influenced by circadian control of isoprene emissions. Nat Geosci 4(10):671–674
14. Häkkinen SAK et al (2013) Semi-empirical parameterization of size-dependent atmospheric nanoparticle growth in continental environments. Atmos Chem Phys 13(15):7665–7682
15. Jimenez JL et al (2009) Evolution of organic aerosols in the atmosphere. Science 326(5959):1525–1529
16. Kirkby J et al (2011) Role of sulphuric acid, ammonia and galactic cosmic rays in atmospheric aerosol nucleation. Nature 476(7361):429–433
17. Kroll JH et al (2006) Secondary organic aerosol formation from isoprene photooxidation. Environ Sci Technol 40(6):1869–1877
18. Kulmala M et al (2013) Direct observations of atmospheric aerosol nucleation. Science 339(6122):943–946

19. Lathière J et al (2005) Past and future changes in biogenic volatile organic compound emissions simulated with a global dynamic vegetation model. Geophys Res Lett 32:L20818
20. Meehl GA et al (2007) Global Climate Projections. In: Climate change 2007: the physical science basis. Solomon S, Qin D, Manning M et al (eds) Contribution of Working Group I to the fourth assessment report of the Intergovernmental Panel on Climate Change. Cambridge University Press, Cambridge
21. Monson RK et al (1992) Relationships among isoprene emission rate, photosynthesis, and isoprene synthase activity as influenced by temperature. Plant Physiol 98:1175–1180
22. Morice CP et al (2012) Quantifying uncertainties in global and regional temperature change using an ensemble of observational estimates: the HadCRUT4 data set. J Geophys Res Atmos 117(D8):D08101
23. Ng NL et al (2007) Effect of NOx level on secondary organic aerosol (SOA) formation from the photooxidation of terpenes. Atmos Chem Phys 7(19):5159–5174
24. Otto A et al (2013) Energy budget constraints on climate response. Nat Geosci 6(6):415–416
25. Paasonen P et al (2013) Warming-induced increase in aerosol number concentration likely to moderate climate change. Nat Geosci 6(6):438–442
26. Pacifico F et al (2012) Sensitivity of biogenic isoprene emissions to past, present, and future environmental conditions and implications for atmospheric chemistry. J Geophys Res 117(D22):D22302
27. Petroff A et al (2008) Aerosol dry deposition on vegetative canopies. Part II: a new modelling approach and applications. Atmos Environ 42(16):3654–3683
28. Petroff A et al (2009) An extended dry deposition model for aerosols onto broadleaf canopies. J Aerosol Sci 40(3):218–240
29. Petroff A, Zhang L (2010) Development and validation of a size-resolved particle dry deposition scheme for application in aerosol transport models. Geosci Model Dev 3(2):753–769
30. Pongratz J et al (2011) Past land use decisions have increased mitigation potential of reforestation. Geophys. Res. Lett. 38(15):L15701
31. Ramanathan V, Xu Y (2010) The Copenhagen Accord for limiting global warming: criteria, constraints, and available avenues. Proc Nat Acad Sci 107(18):8055–8062
32. Rosenstiel TN et al (2003) Increased $CO_2$ uncouples growth from isoprene emission in an agriforest ecosystem. Nature 421(6920):256–259
33. Schmidt A et al (2012) Importance of tropospheric volcanic aerosol for indirect radiative forcing of climate. Atmos Chem Phys 12(16):7321–7339
34. Spracklen DV et al (2012) Observations of increased tropical rainfall preceded by air passage over forests. Nature 489(7415):282–285
35. Tunved P et al (2008) The natural aerosol over Northern Europe and its relation to anthropogenic emissions—implications of important climate feedbacks. Tellus B 60(4):473–484
36. Williams JE et al (2013) Quantifying the uncertainty in simulating global tropospheric composition due to the variability in global emission estimates of Biogenic Volatile Organic Compounds. Atmos Chem Phys 13(5):2857–2891
37. Yli-Juuti T et al (2011) Growth rates of nucleation mode particles in Hyytiälä during 2003–2009: variation with particle size, season, data analysis method and ambient conditions. Atmos Chem Phys 11(24):12865–12886
38. Zeng Z et al. (2013) Committed changes in tropical tree cover under the projected 21st century climate change. Sci Rep 3:1951

Zeitfracht Medien GmbH
Ferdinand-Jühlke-Straße 7
99095 Erfurt, Deutschland
produktsicherheit@kolibri360.de